T0183904

Communications
in Computer and Information Science   862

Commenced Publication in 2007
Founding and Former Series Editors:
Phoebe Chen, Alfredo Cuzzocrea, Xiaoyong Du, Orhun Kara, Ting Liu,
Krishna M. Sivalingam, Dominik Ślęzak, Takashi Washio, and Xiaokang Yang

More information about this series at http://www.springer.com/series/7899

Christoph Quix · Jorge Bernardino (Eds.)

Data Management Technologies and Applications

7th International Conference, DATA 2018
Porto, Portugal, July 26–28, 2018
Revised Selected Papers

 Springer

Editors
Christoph Quix
RWTH Aachen University
Aachen, Nordrhein-Westfalen, Germany

Jorge Bernardino
University of Coimbra
Coimbra, Portugal

ISSN 1865-0929   ISSN 1865-0937   (electronic)
Communications in Computer and Information Science
ISBN 978-3-030-26635-6   ISBN 978-3-030-26636-3   (eBook)
https://doi.org/10.1007/978-3-030-26636-3

This Springer imprint is published by the registered company Springer Nature Switzerland AG
The registered company address is: Gewerbestrasse 11, 6330 Cham, Switzerland

Preface

The present book includes extended and revised versions of a set of selected papers from the 7th International Conference on Data Science, Technology and Applications (DATA 2018), held in Porto, Portugal, during July 26–28, 2018.

DATA 2018 received 69 paper submissions from 26 countries, of which 13% are included in this book. The papers were selected by the event chairs and their selection is based on a number of criteria that include the classifications and comments provided by the Program Committee members, the session chairs' assessment, and also the program chairs' global view of all papers included in the technical program. The authors of selected papers were then invited to submit a revised and extended version of their papers having at least 30% innovative material.

"Value of Data in a Data-Driven Economy" was the theme of the 7th International Conference on Data Science, Technology and Applications (DATA), whose purpose was to bring together researchers, engineers, and practitioners interested in databases, big data, data mining, data management, data security and other aspects of information systems and technology involving advanced applications of data.

The nine papers selected to be included in this book contribute to the understanding of relevant trends of current research on data science, technology and applications. There are three papers addressing the topic of databases and data security focusing on query processing and optimization, database architecture and performance, architectural concepts, data privacy and security, big data as a service, and NoSQL databases. Another three papers address the topic of soft computing in data science focusing on data mining, semi-structured and unstructured data, text analytics, data science, neural network applications, predictive modelling, evolutionary computing and optimization, hybrid methods, and deep learning. There are two papers on the topic of business analytics, focusing on datamining, data analytics, predictive modelling, industry 4.0, decision support systems, and statistics exploratory data analysis. Finally, we have one paper approaching the important aspect of data management and quality including data modelling and visualization, information visualization, and data science.

We would like to thank all the authors for their contributions and also to the reviewers who helped ensure the quality of this publication.

July 2018

Christoph Quix
Jorge Bernardino

Organization

Conference Chair

Jorge Bernardino — Polytechnic Institute of Coimbra - ISEC, Portugal

Program Chair

Christoph Quix — Hochschule Niederrhein, University of Applied Sciences and Fraunhofer, Germany

Program Committee

Stefan Conrad	Heinrich-Heine University Düsseldorf, Germany
James Abello	Rutgers, The State University of New Jersey, USA
Muhammad Abulaish	South Asian University, India
Hamideh Afsarmanesh	University of Amsterdam, The Netherlands
Christos Anagnostopoulos	University of Glasgow, UK
Nicolas Anciaux	Inria Paris-Rocquencourt, France
Bernhard Bauer	University of Augsburg, Germany
Fevzi Belli	Izmir Institute of Technology, Turkey
Karim Benouaret	Université Claude Bernard Lyon 1, France
Nikos Bikakis	University of Ioannina, Greece
Christian Bizer	University of Mannheim, Germany
Gloria Bordogna	National Research Council, Italy
Francesco Buccafurri	University of Reggio Calabria, Italy
Cinzia Cappiello	Politecnico di Milano, Italy
Sudarshan Chawathe	University of Maine, USA
Ruey Chen	Cute, Taiwan
Gianni Costa	ICAR-CNR, Italy
Theodore Dalamagas	Athena Research and Innovation Center, Greece
Bruno Defude	Institut Mines Telecom, France
Steven Demurjian	University of Connecticut, USA
Fabien Duchateau	Université Claude Bernard Lyon 1/LIRIS, France
John Easton	University of Birmingham, UK
Todd Eavis	Concordia University, Canada
Neamat El Tazi	Cairo University, Egypt
Mohamed Eltabakh	Worcester Polytechnic Institute, USA
Markus Endres	University of Augsburg, Germany
João Ferreira	ISCTE, Portugal
Gustavo Figueroa	Instituto Nacional de Electricidad y Energías Limpias, Mexico
Sergio Firmenich	Universidad Nacional de La Plata, Argentina

Marco Villani   University of Modena and Reggio Emilia, Italy
Gianluigi Viscusi   EPFL Lausanne, Switzerland
Hannes Voigt   Neo4j, Germany
Zeev Volkovich   Ort Braude College, Israel
George Vouros   University of Piraeus, Greece
Fan Wang   Microsoft, USA
Yun Xiong   Fudan University, China
Filip Zavoral   Charles University Prague, Czech Republic
José-Luis Zechinelli-Martini   Universidad de las Americas Puebla, Mexico
Jiakui Zhao   State Grid Information and Telecommunication Group
                                    of China, China

Additional Reviewers

Georgios Chatzigeorgakidis   University of Peloponnese, Greece
Pantelis Chronis   University of Peloponnese, Greece
Iván García Miranda   University of Valladolid, Spain
Ludmila Himmelspach   Heinrich-Heine-University Duesseldorf, Germany
Gerhard Klassen   Heinrich-Heine Universität Düsseldorf, Germany
Vimal Kunnummel   University of Vienna, Austria

Invited Speakers

Carsten Binnig   TU Darmstadt, Germany
Tova Milo   Tel Aviv University, Israel

Contents

Constructing a Data Visualization Recommender System

Petra Kubernátová[1]($\boxtimes$), Magda Friedjungová[2], and Max van Duijn[1]

[1] Leiden Institute of Advanced Computer Science,
Leiden University, Leiden, The Netherlands
pkubernatova@gmail.com, m.j.van.duijn@liacs.leidenuniv.nl
[2] Faculty of Information Technology,
Czech Technical University in Prague, Prague, Czech Republic
magda.friedjungova@fit.cvut.cz

Abstract. Choosing a suitable visualization for data is a difficult task. Current data visualization recommender systems exist to aid in choosing a visualization, yet suffer from issues such as low accessibility and indecisiveness. In this study, we first define a step-by-step guide on how to build a data visualization recommender system. We then use this guide to create a model for a data visualization recommender system for non-experts that aims to resolve the issues of current solutions. The result is a question-based model that uses a decision tree and a data visualization classification hierarchy in order to recommend a visualization. Furthermore, it incorporates both task-driven and data characteristics-driven perspectives, whereas existing solutions seem to either convolute these or focus on one of the two exclusively. Based on testing against existing solutions, it is shown that the new model reaches similar results while being simpler, clearer, more versatile, extendable and transparent. The presented guide can be used as a manual for anyone building a data visualization recommender system. The resulting model can be applied in the development of new data visualization software or as part of a learning tool.

Keywords: Data visualization · Recommender system · Non-expert users

1 Introduction

In recent years, we have been witnesses to the rise of big data as one of the key topics of computer science. With huge amounts of data being generated every second of every day, it has become a necessity to focus on its storage, analysis and presentation. Data visualization is aiding us in the presentation phase. By definition, it is the representation of information in a visual form, such as a chart, diagram or picture. It can find its place in a variety of areas such as art, marketing, social relations and scientific research. There were over 300 visualization types available at the time of writing this paper [1]. This growing number

© Springer Nature Switzerland AG 2019
C. Quix and J. Bernardino (Eds.): DATA 2018, CCIS 862, pp. 1–25, 2019.
https://doi.org/10.1007/978-3-030-26636-3_1

makes the choice of choosing a suitable data visualization very difficult. Especially if one is not particularly skilled in the area. Thankfully, data visualization recommender systems exist to help us with this complicated task.

This paper is an extended version of [2] which was presented at the DATA 2018 conference. The main contribution of this extended version lies in including a step-by-step guide for building a data visualization recommender system. We also present a more extensive existing solution study. We further clarify the process of constructing our model and also elaborate on ways of evaluating and implementing it. Section 2 places data visualization recommender systems in the context of data science. Section 3 introduces our step-by-step guide to building a data visualization recommender system. In Sects. 4, 5, 6, 7, 8, 9 and 10 we go through the individual steps and build our very own data visualization recommender system while taking measures to make it well suited for non-expert users. We define a 'non-expert user' as someone without professional or specialized knowledge of data visualization. We thus include both complete beginners and users who have general knowledge of data visualization types (e.g. bar charts, pie charts, scatter plots) but have no professional experience in the fields of data science and data communication. We want to see if we can make adjustments that make a system more suitable for non-expert users while maintaining effectiveness (still clearly distinguishing the data visualizations from each other) and performance (recommending the most suitable visualization type). We draw conclusions in Sect. 11 and set an agenda for future work in Sect. 12.

2   Context

2.1   Data Science

Data science plays an important role in scientific research, as it aids us in collecting, organizing, and interpreting data, so that it can be transformed into valuable knowledge. Figure 1 shows a simplified diagram of the data science process as described by O'Neil and Schutt [3]. This diagram is helpful in demarcating the research objectives of this paper. According to O'Neil and Schutt, first, real world raw data is collected, processed and cleaned through a process called data munging. Then exploratory data analysis (EDA) follows, during which we might find that we need to collect more data or dedicate more time to cleaning and organizing the current dataset. When finished with EDA, we may use machine learning

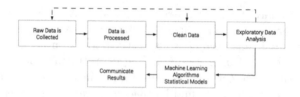

Fig. 1. The data science process [3].

algorithms, statistical models and data visualization techniques, depending on the type of problem we are trying to solve. Finally, results can be communicated [3].

Our focus here is on the part of the process concerning exploratory data analysis or EDA. EDA uses a variety of statistical techniques, principles of machine learning, but also, crucially, the data visualization techniques we study in this paper. Please note that data visualization can also be a part of the "Communicate Results" stage of the data science process. There is a thin line between data visualizations made for exploration and ones made for explanation, as most exploratory data visualizations also contain some level of explanation and vice-versa.

2.2 Exploratory Data Analysis

Exploratory data analysis (EDA) is not only a critical part of the data science process, it is also a kind of philosophy. You are aiming to understand the data and its shape and connect your understanding of the process that collected the data with the data itself. EDA helps with suggesting hypotheses to test, evaluating the quality of the data, identifying potential need for further collection or cleaning, supporting the selection of appropriate models and techniques and, most importantly for the context of this study, it helps find interesting insights in your data [4].

2.3 Data Visualization

There are many definitions of the term data visualization. The one used in this study is: data visualization is the representation and presentation of data to facilitate understanding. According to Kirk, our eye and mind are not equipped to easily translate the textual and numeric values of raw data into quantitative and qualitative meaning. "We can look at the data, but we cannot understand it. To truly understand the data, we need to see it in a different kind of form. A visual form" [5].

Illinsky and Steele describe data visualization as a very powerful tool for identifying patterns, communicating relationships and meaning, inspiring new questions, identifying sub-problems, identifying trends and outliers, discovering or searching for interesting or specific data points [6].

Munzner made a 3-step model for data visualization design. According to this model, we first need to decide what we want to show. Secondly, we need to motivate why we want to show it. Finally, we need to decide how we are going to show it [7]. There are many different types of data visualizations to help us with the third step. However, the challenge remains in choosing the most suitable one. Data visualization recommender systems were made to help with this difficult task. We find that the WHAT and the WHY greatly influence the HOW, thus we aim to build a system that reflects all three aspects of the data visualization design process in some way.

2.4   Data Visualization Recommender Systems

Within this study we define data visualization recommender systems as tools that seek to recommend visualizations which highlight features of interest in data. This definition is based on combining common aspects of definitions in existing work.

While the output of data visualization recommender systems is always a recommendation for data visualization types in some shape or form, the input can differ. It can be, for example, just the data itself, a specification of goals or the specification of aesthetic preferences. The type of input affects the type of recommendation strategy used and consequently the type of the recommender system. Kaur and Owonibi distinguish 4 types of recommender systems [8]:

– **Data Characteristics Oriented.** These systems recommend visualizations based on data characteristics.
– **Task Oriented:** These systems recommend visualizations based on representational goals as well as data characteristics.
– **Domain Knowledge Oriented.** These systems improve the visualization recommendation process with domain knowledge.
– **User Preferences Oriented.** These systems gather information about the user presentation goals and preferences through user interaction with the visualization system.

The line between different categories of recommendation systems is rather thin and some systems can have ambiguous classifications, as will be discussed.

3   Building a Data Visualization Recommender System

3.1   Previous Work Overview

In any type of research, it is important to explore the state of the art of the corresponding field first. We will examine the area of data visualization recommender systems and its history. We are going to look at current solutions and identify which aspects could be used in our own system. Secondarily, the previous work overview will also help us determine which solutions are most suitable to be used while testing our system later on.

3.2   Exploratory Survey

Next, it is crucial to get the potential users involved. We will create and run an exploratory survey among different data science communities on Facebook and LinkedIn. We focus on data science communities, because it will ensure that our respondents will have some sort of familiarity with data science and its terminology. We made this choice for convenience and time management reasons. The survey should aid us in making decisions about our system. As we want to explore options of making the system easy to use for non-experts, it is important to have a clear definition of a non-expert. The survey will help us in this specification as well.

3.3 Forming Requirements

The results of both the previous work overview and the survey will help us formulate requirements which our system should fulfill. These will serve as good evaluation criteria for our system later on.

3.4 Constructing a Model

Once the requirements are formulated, we will begin constructing a model for our recommender system. First we will have to decide which form the model will take - what the base structure will be. We will then establish the different components of the structure. Finally, we will combine it all together.

3.5 Testing the Model

At this point we will take time to test the model that we have constructed. We will perform two tests. The first one will focus on determining whether the model is able to produce results similar to existing solutions. The second test focuses on testing the extensibility of the model by adding a new type of visualization.

3.6 Evaluating the Model

We will evaluate the model according to the results of the tests and also comment on how it manages to fulfill the requirements we have set. We will most probably have to make adjustments to the model.

3.7 Implementing the Model

Finally, we will implement the model we have created and make it come to life as a data visualization recommender system.

4 Previous Work Overview

While performing the existing solutions study, we have discovered that the amount of works in the area focusing on domain knowledge oriented and user preferences oriented data visualization recommender systems is not sufficient. We made the decision to focus on data characteristics oriented and task oriented systems. As we stated earlier, there is a thin line between the different types of recommendation systems, so in some solutions you might still see elements of the two types of recommender systems we do not explicitly focus on.

4.1 Data Characteristics Oriented Systems

Systems based on data characteristics aim to improve the understanding of the data, of different relationships that exist within the data and of procedures to represent them. Some of the following tools and techniques are not recommendation systems per se but they were a crucial part of the history of this field and foundations for other recommender systems stated, thus we feel it is appropriate to list them as well.

BHARAT was the first system that proposed rules for determining which type of visualization is appropriate for certain data attributes [9]. As this work was written in 1981, the set of possible visualizations was not as varied as it is today. The system incorporated only the line, pie and bar charts and was based on a very simple design algorithm. If the function was continuous, a line chart was recommended. If the user indicated that the range sets could be summed up to a meaningful total, a pie chart was recommended and bar charts were recommended in all the remaining cases. Even though this system would now be considered very basic, it served as the foundation for other systems that followed.

APT. In 1986, Mackinlay proposed to formalize and codify the graphical design specification to automate the graphics generation process [10]. His work is based on the work of Joseph Bertin, who came up with a semiology of graphics, where he specified visual variables such as position, size, value, color, orientation etc. and classified them according to which features they communicate best. For example, the shape variable is best used to show differences and similarities between objects. Mackinlay codified Bertin's semiology into algebraic operators that were used to search for effective presentations of information. He based his findings on the principals of expressiveness and effectiveness. Expressiveness is the idea that graphical presentations are actually sentences of graphical languages that have precise syntactic and semantic definitions, while effectiveness refers to how accurately these presentations are perceived. He aimed to develop a list of graphical languages that can be filtered with the expressiveness criteria and ordered with the effectiveness criteria for each input. He would take the encoding technique and formalize it with primitive graphical language (which data visualizations can show this), then he would order these primitive graphical languages using the effectiveness principle (how accurately perceived they are). APT's architecture was focused on how to communicate graphically rather than on what to say. Casner extended this work by comparing design alternatives via a measure of the work that was required to read presentations, depending on the task [11]. Roth and Mattis added additional types of visualizations [12].

VizQL (Visual Query Language). In 2003, Hanrahan revised Mackinglay's specifications into a declarative visual language known as VizQL [13]. It is a formal language for describing tables, charts, graphs, maps, time series and tables of visualizations. The language is capable of translating actions into a database query and then expressing the response graphically.

Tableau and Its Show Me Feature. The introduction of Tableau was a real milestone in the world of data visualization tools. Due to the simple user interface, even inexperienced users could create data visualizations. It was created when Stolte, together with Hanrahan and Chabot, decided to commercialize a system called Polaris [14] under the name Tableau Software. In 2007 Tableau introduced a feature called Show Me [15]. The Show Me functionality takes advantage of VizQL to automatically present data. At the heart of this feature is a data characteristics-oriented recommendation system. The user selects the data attributes that interest him and Tableau recommends a suitable visualization. Tableau determines the proper visualization type to use by looking at the types of attributes in the data. Each visualization requires specific attribute types to be present before it can be recommended. For example, a scatter plot requires 2 to 4 quantitative attributes. Furthermore, the system also ranks every visualization on familiarity and design best practices. Finally, it recommends the highest-ranked eligible visualization. Mackinlay and his team have also performed interesting user tests with the Show Me feature. They found that the Show Me feature is being used (very) modestly by skilled users (i.e. in only 5.6% of cases).

Tableau inspired us by it's simple user interface which is suitable for non-experts, reminding us that our model should enable a simple user interface implementation. Furthermore, we make use of their classification of data visualizations based on design best practices and familiarity as well as the conditions that the data must fulfill for a specific data visualization to be chosen. The fact that Tableau is so widely used and that a demo version is freely accessible determined it suitable for use in our tests.

ManyEyes. Viegas et al. created the first known public website where users may upload data and create interactive visualizations collaboratively: ManyEyes [16]. The tool was created for non-experts, as Viegas et al. wanted to make a tool that was accessible for anyone regardless of prior knowledge and training. Design choices were made to reflect the effort to find a balance between powerful data-analysis capabilities and accessibility to the non-expert visualization user. The visualizations were created by matching a dataset with one of the 13 types of data visualizations implemented in the tool. To set up this matching, the visualization components needed to be able to express its data needs in a precise manner. They divided the data visualizations into groups by data schemas. A data schema could be, for example, "single column textual data". Thus, a bar chart was described as "single column textual data and more than one numerical value". The dataset and produced visualization could then be shared with others for comments, feedback and improvement [16]. However, the tool closed down in 2015.

ManyEyes taught us that the way to attract non-expert users is to make the application resulting from our model as accessible as possible. This means that our model could be suited to web-based implementations.

Watson Analytics. Since 2014, IBM have been developing a tool called Watson Analytics [17]. It carries the same name as another successful IBM project - the Watson supercomputer, which combines artificial intelligence and sophisticated analytical software to perform as a "question-answering" system. In 2011, it famously defeated top-ranked players in a game of Jeopardy!. Similarly to the Watson supercomputer, Watson Analytics uses principles of machine learning and natural language processing to recommend users either questions they can ask about their data, or a specific visualization. However, IBM has not revealed what values or attributes are used by the recommendation system to select a visualization.

Watson Analytics reminded us that the structure of our model should be variable enough to be suitable for implementing machine learning and artificial intelligence techniques on it for the model to possibly improve itself. A demo version of the system is freely available, so we use it in our tests.

VizDeck. In 2012 Key et al. developed a tool called VizDeck [18]. The web-based tool recommends visualizations based on statistical properties of the data. It adopts a card game metaphor to organize multiple visualizations into an interactive visual dashboard application. Vizdeck was created as Key et al. found that scientists were not able to self-train quickly in more sophisticated tools such as Tableau. The tool supports scatter plots, histograms, bar charts, pie charts, timeseries plots, line plots and maps.

Based on the statistical properties of the underlying dataset, VizDeck generates a "hand" of ranked visualizations and the user chooses which "cards" to keep and put into a dashboard and which to discard. Through this, the system learns which visualizations are appropriate for a given dataset and improves the quality of the "hand" dealt to future users. For the actual recommendation system part of the tool, they trained a model of visualization quality that relates statistical features of the dataset to particular visualizations. As far as we know VizDeck was never actually deployed and remained at the testing phase.

VizDeck again inspired us to think about the possibility of our model being self-improving and educative.

Microsoft Excel's Recommended Charts Feature. In the 2013 release of Microsoft Excel, a new feature called Recommended Charts was introduced. The user can select the data they want to visualize and Excel recommends a suitable visualization [19]. However, Microsoft does not share exactly how this process is carried out, making it less suitable as a source of inspiration. We use Microsoft Excel to test our model, because it is accessible.

SEEDB. In 2015 Vartak et al. proposed an engine called SEEDB [20]. They judge the interestingness of a visualization based on the following theory: a visualization is likely to be interesting if it displays large deviations from some reference (e.g. another dataset, historical data, or the rest of the data). This helps them identify the most interesting visualizations from a large set of potential visualizations. They identified that there are more aspects that determine the interestingness of a visualization, such as aesthetics, user preference, metadata and user tasks. A full-fledged visualization recommendation system should take into account a combination of these aspects. A major disadvantage of SEEDB is that it only uses variations of bar charts and line charts. As far as we know SEEDB was never deployed.

SEEDB made us think about having multiple views in our model from different interestingness perspectives, because we want our model to be full-fledged, as they describe.

Voyager. In 2016, Wongsuphasawat et al. developed a visualization recommendation web application called Voyager [21], based on the Compass recommendation engine [22] and a high-level specification language called Vega-lite [23]. It couples browsing with visualization recommendation to support exploration of multivariate, tabular data. First, Compass selects variables by taking user-selected variable sets and suggesting additional variables. It then applies data transformations (e.g. aggregation or binning) to produce a set of derived data tables. For each data table, it designs encodings based on expressiveness and effectiveness criteria and prunes visually similar results. The user then includes or excludes different variables to focus on a particular set of variables that are interesting. Voyager is a tool which is freely available online, which makes it suitable for use in our tests.

Google Sheets. Google Sheets [24] is a tool which allows users to create, edit and share spreadsheets. It was introduced in 2007 and is very similar to Microsoft Excel. In June of 2017, the tool was extended with the Explore Feature, which helps with automatic chart building and data visualization. It uses elements of artificial intelligence and natural language processing to recommend users questions they might want to ask about their data, as well as recommending data visualizations that best suit their data. In the documentation for this feature, Google specifies each of the included data visualizations using functions and conditions that have to be fulfilled in order for that particular data visualization to be recommended. However, it does not reveal exactly how it chooses the most suitable data visualization, because a couple of visualizations have the same conditions. We make use of the classification of data visualizations presented in Google Sheets and thanks to its accessibility online, we use it in our tests.

4.2 Task Oriented Systems

Task-oriented systems aim to design different techniques to infer the representational goal or a user's intentions. In 1990, Roth and Mattis were the first to identify different domain-independent information seeking goals, such as comparison, distribution, correlation etc. [12]. Also in 1990, Wehrend and Lewis proposed a classification scheme based on sets of representational goals [25]. It was in the form of a 2D matrix where the columns were data attributes, the rows representational goals and the cells data visualizations. To find a visualization, the user had to divide the problem into subproblems, until for each subproblem it was possible to find an entry in the matrix. A representation for the original complex problem could then be found by combining the candidate representation methods for the subproblems. Unfortunately, the complete matrix was not published so it is unknown which specific types of data visualizations were included.

BOZ. BOZ is an automated graphic design and presentation tool that designs graphics based on an analysis of the task which a graphic is intended to support [11]. The system analyzes a logical description of a task to be performed and designs an equivalent perceptual task. BOZ produces a graphic along with a perceptual procedure describing how to use the graphic to complete the task. It is able to design different presentations of the same information customized to the requirements of different tasks.

The BOZ system reminded us that the difference between a suitable and non-suitable data visualization could also lie in the way that humans perceive them. For example, a pie chart is generally considered not suitable, as humans have difficulty judging the size of angles accurately. A bar chart is more suitable for the task.

IMPROVISE. In the previous studies, the user task list was manually created. However, in 1998, Zhou and Feiner introduced advanced linguistic techniques to automate the derivation of the user task from a natural language query [26]. They introduced a visual task taxonomy to automate the process of gaining presentation intents from the text. The taxonomy interfaces between high level tasks that can be accomplished by low level visualization techniques. For example, the visual task "Focus" implies that visual techniques such as "Enlarge" or "Highlight" could be used. This taxonomy is implemented in IMPROVISE.

An example of an IMPROVISE use case is presenting an overview of a hospital patient's information to a nurse. To achieve this goal, it constructs a structure diagram that organizes various information (e.g. IV lines) around a core component (the patient's body). In a top-down design manner, IMPROVISE first creates an 'empty' structure diagram and then populates it with components by partitioning and encoding the patient information into different groups.

HARVEST. In 2009, Gotz and Wen introduced a novel behavior-driven approach [27]. Instead of needing explicit task descriptions, they use implicit task information obtained by monitoring users' behavior to make recommendation more effective. The Behavior-Driven Visualization Recommendation (BVDR) approach has two phases. In the first phase of BDVR, they detect four predefined patterns from user activity. In the second phase, they feed the detected patterns into a recommendation algorithm, which infers user intent in terms of common visual tasks (e.g. comparison) and suggests visualizations that better support the user's needs. The inferred visual task is used together with the properties of the data to retrieve a list of potentially useful visual metaphors from a visualization example corpus made by Zhou and Chen [28]. It contains over 300 examples from a wide variety of sources. Unfortunately, we were not able to access this corpus.

The conclusions made from HARVEST gave us the idea to provide explanations why a certain data visualization was recommended to enhance the educative aspect of our model.

DataSlicer. A recent study by Alborzi et al. takes yet another approach [29]. The authors' hypothesis is that for many data sets and common analysis tasks, there are relatively few "data slices" that result in effective visualizations. Data slices are different subsets of data. Their objective is to improve the user experience by suggesting data slices that, when visualized, present correct solutions to the user's task in an effective way. At any given time in working on the task, users may ask the system to suggest visualizations that would be useful for solving the task. A data slice is considered interesting if past users spent a considerable amount of time looking at its visualization. They first developed a framework which captures exemplary data slices for a user task, explores and parses visual-exploration sequences into a format that makes them distinct and easy to compare. Then they developed a recommendation system, DataSlicer, that matches a "currently viewed" data slice with the most promising "next effective" data slices for the given exploration task. In user tests, DataSlicer significantly improved both the accuracy and speed for identifying spatial outliers, data outliers, outlier patterns and general trends. The system quickly predicted what a participant was searching for based on their initial operations, then presented recommendations that allowed the participants to transform the data, leading them to desired solutions.

The system is interesting, because it deals with the problem of efficiently leading casual or inexperienced users to visualizations of the data that summarize in an effective and prominent way the data points of interest for the user's exploratory-analysis task. The authors do not specify exactly which tasks they include in their system.

All in all, we identify some pitfalls of the existing systems. Such as them not being accessible enough, too complicated, too formal and too secretive when it comes to their recommendation process. The biggest pitfall is that the result of their recommendation process is most commonly a set of data visualizations, which, in our opinion, leaves the users a bit further than they started, but still nowhere, because they still have to choose the most suitable visualization. The possibilities have been narrowed, but a decision still must be made. We hope to avoid these pitfalls within our model.

We establish that we are going to test our model against the solutions available to us. This means Tableau, Watson Analytics, Excel, Voyager and Google Sheets. Please note that we are going to compare against the recommendation system features of the tools, not the tools as a whole.

5   Exploratory Survey

5.1   Participants

In total, we gathered 88 valid responses ($n = 88$). Out of the 88 respondents, 78% ($n = 69$) were male and 22% ($n = 19$) female. The average age was 29.86 years. We asked the respondents to indicate their knowledge level on a scale of 1 to 10, 1 being beginner and 10 being expert. The average knowledge level was 5.70. We opted to divide the scale into three ranges in the following way:

1–3 are beginners, 4–7 are non-experts and 8–10 are experts. According to our ranges we had 26% $(n = 23)$ beginner level, 44% non-expert $(n = 39)$ level and 30% $(n = 26)$ expert level respondents.

5.2   Questions

Questions in our survey included:

- How long have you been working in a data visualization related field?
- Which software do you mostly use to create your data visualizations?
- What are your top 3 most used visualization techniques? (e.g. bar chart, scatterplot, line chart, treemap...)
- What is your main goal when you make data visualizations?
- Which of the following tasks do you usually perform using your data visualization?
- Do you know any data visualization recommender systems?

5.3   Results

The results of our survey can be summarised as follows:

- For all groups, the main purpose of making data visualizations was for analysis (65% of beginners, 64% of non-experts, 58% of experts).
- All types of users choose data visualizations mainly according to: the characteristics of their data (57% of beginners, 62% of non-experts, 65% of experts) and the tasks that they want to perform (48% of beginners, 51% of non-experts, 62% of experts).
- For all groups, the two most used visualizations are bar charts (17% of beginners, 38% of non-experts, 35% of experts) and scatter plots (43% of beginners, 26% of non-experts, 31% of experts).
- All groups were mostly unable to name an existing data visualization recommendation system (0% able vs. 100% unable for beginners, 5% able vs. 95% unable for non-experts and 4% able vs. 96% unable for experts).
- All groups would be willing to use a data visualization recommendation system, although experts were less willing than beginners and non-experts (100% willing vs. 0% not willing for beginners, 87% willing vs. 13% not willing for non-experts and 77% willing vs. 23% not willing for experts).

The most crucial finding that we made from the results of our survey was, that the approaches of beginners, non-expert and expert users do not differ enough for us to be able to clearly determine what specific features a non-expert user would need in a data visualization recommender system model. This was an unexpected result for us.

6   Forming Requirements

Based on research of previous approaches to our problem and the results of our survey, we have identified the following requirements which our model should fulfill:

1. **Simplicity** - The model should be simple, it must have good flow and a very straightforward base structure.
2. **Clarity** - We aim for the result of our recommendation system to be one data visualization. Not a set, like in some current tools. This means that the underlying classification hierarchy of data visualizations must be clear and unambiguous.
3. **Versatility** - We want our model to combine different kinds of recommendation systems. From our survey we learn that when users select a suitable data visualization type, they do so based on the characteristics of their data and the tasks they want to perform. Based on this we incorporate a data characteristics-oriented and task-oriented approach. Furthermore, we want our model to be easily implemented in different programming languages and environments.
4. **Extendibility** - Our aim is for our model to be easily extendable. We want the process of adding visualizations into the model to be simple. We want it to be a useful "skeleton" which can be easily extended to include automatic visualizations etc.
5. **Education** - We want our model to not only function as a recommender system, but also as a learning tool.
6. **Transparency** - Once we recommend a visualization, we want the users to see, why that particular visualization was recommended, meaning that the path to a visualization recommendation through our model has to be retraceable.
7. **Self-learning** - We want our model to be able to improve itself. This means, amongst other things, that it should be machine learning friendly.
8. **Competitiveness** - We want our model to still produce results which are comparable to results from other systems.

7   Constructing a Model

7.1   Base Structure

Since the aim of our model is to help a user *decide* which data visualization to use, the obvious choice seemed to be to use the structure of decision trees. A decision tree has four main parts: a root node, internal nodes, leaf nodes and branches. Decision trees can help uncover unknown alternative solutions to a problem and they are well suited for machine learning methods.

Once we determined that the decision tree was a possible base structure, we needed to specify what our root node, internal nodes, leaf nodes and branches would be. It was clear right away that the leaf nodes would be the different types of data visualizations since that was the outcome that we wanted to achieve. The root node, internal nodes and branches are inspired by a question-based game that you might have played in your childhood called "21 Questions". In this game, one player thinks of a character and the other players must guess who it is using at most 21 questions. The questions can only be answered 'yes' or 'no'. In our instance, the character the player is thinking of is the leaf node (data visualization), the questions are both internal nodes and the root node and the 'yes' and 'no' answers are branches.

7.2   Root and Internal Nodes

As previously stated, the internal nodes and root node of our model are questions. Just like the "21 Questions" game, each question should eliminate a lot of possible characters (data visualizations). At the same time, the questions have to be understandable (even for non-experts). It became clear that we had to construct questions, which would possess the ability to clearly distinguish different types of data visualizations. The subjects of these questions must then be features that distinguish the different data visualizations from each other. We call these features "distinguishing features". The key to solving this problem is to base the questions on a clear classification hierarchy. As far as we know, there is no one specific classification hierarchy of data visualizations which would be used globally. We had to put together a classification of our own. We decided to research different methods of classification and combine them together. This was a very time-consuming process. We went through a total of 19 books [3,5–7,9,31–44] and for each one, we constructed a diagram representing the classification that was described in the text. We considered the possibility of automating this process, but that would provide enough material to cover a whole other paper.

We examined the classification hierarchies from books together with hierarchies available from web resources and existing tools. We also made note of any advantages or disadvantages of a specific data visualization, if they were listed. For example in several sources [3,5,6] the authors stated that the pie chart is not suitable for when you have more than 7 parts. The advantages and disadvantages reflected features of the data visualizations that could determine whether they are candidates for recommendation or not, so they are key for the final model.

We identified that there are two basic views that the classifications incorporate. The first one is a view from the perspective of the task the user wants to perform. The second is a view from the perspective of the characteristics of the data the user has available. This is in line with data characteristics and task oriented recommendation systems [8].

We have identified a prominent issue in the classification hierarchies: they mix different views into one without making a clear distinction between them. To avoid this issue, we have selected the root node of our model to be a question which would distinguish between two views. The first view is from a task-based perspective and it uses the representational goal or user's intentions behind visualizing the data to recommend a suitable visualization. The second view is from a data-driven perspective, where a visualization recommendation is made based on gathering information about the user's data. The root node of our model is a question asking "Do you know what your main task is?" If the user answers "Yes", he is taken in to the task-based branch. If he answers "No", he is taken straight into the data characteristics-based branch.

Once we established the root node, we had to specify the internal nodes. Based on the findings we made in previous paragraphs, we have established a list of distinguishing features and their hierarchy. We present a part of the hierarchy as an example:

- Suitability for a specific task
  - Comparing
    * Over time
    * Quantities
    * Proportions
    * Other
  - Analyzing
    * Trends
    * Correlations
    * Distribution
    * Patterns
    * Clusters

Based on the distinguishing features, we have constructed questions that ask whether that feature is present or not. For example, for the distinguishing features stated above:

1. Is your main task to compare over time?
2. Is your main task to compare quantities?
3. Is your main task to compare proportions?
4. Is your main task to compare something else?
5. Is your main task to analyze trends?
6. Is your main task to analyze correlations?
7. Is your main task to analyze distribution?
8. Is your main task to analyze patterns?
9. Is your main task to analyze clusters?

7.3 Leaf Nodes

The main challenge in this part of the process was to decide which of the more than 300 types of data visualizations available [1] to include in the initial version of our model. We took a rather quantitative approach to the problem. We went through all the different classification hierarchies we constructed previously and extracted a list of the data visualizations that occur. We removed duplicates (different names for the same visualization, different layouts of the same visualization) and we counted how many times each data visualization occurred. The ones that occurred 5 times or more were included in our final model. The final list contains 29 data visualizations. Since one of our requirements for the final model is easy extendibility, we feel that 29 data visualizations are appropriate for the initial model. The list includes: Bar Chart, Pie Chart, Bubble Chart, Cartogram, Radar Plot, Scatter Plot, Scatter Plot Matrix, Slope Graph, Heat Map, Histogram, Line Chart, Tree Map, Network, Stacked Line Chart etc.

7.4   Final Model

We classified each of our leaf nodes (data visualizations) using the distinguishing features we constructed previously. For each data visualization, we answered the set of questions relating to the distinguishing features. This revealed how the questions have to be answered in order to get to a certain data visualization.

We then combined all the classifications together to construct the final model[1]. The model has 107 internal nodes and 105 leaf nodes. It always results in a recommendation. If no other suitable visualization is found, we recommend to use a table by default. Other data visualization recommender systems such as Tableau adopt this behavior as well.

8   Testing the Model

8.1   Competitiveness Test

The aim of this test was to determine whether our model was able to compete with existing systems in terms of similarity of solutions. We obtained 10 different test data sets with various features (See Table 1). The data sets were preprocessed to remove invalid entries and to ensure that all the attributes were of the correct data type.

For each data set, we formulated an example question that a potential user is aiming to answer. This was done in order to determine which attributes of the data would be used in the recommendation procedure. Most existing tools require the user to select the specific attributes that they want to use for their data visualization. By specifying these for each data set we attempt to mimic this behavior. Table 1 shows the data sets along with their descriptions.

We named our model **NEViM**, which stands for Non-Expert Visualization Model. We tested our model against existing solutions which are freely available: Tableau (10.1.1), Watson Analytics (version available in July 2017), Microsoft Excel (15.28 Mac), Voyager (2) and Google Sheets (version available in July 2017). For each system and every data set, we aimed to achieve a recommendation for a data visualization that would answer the question and incorporate all the specified attributes in one graph as there is no possible way to answer the question without incorporating the specified attributes. Some systems solve more complex questions by creating a series of different data visualizations, with each visualization incorporating a different combination of attributes. We excluded

---

[1] The whole model can be viewed at a website dedicated to this research project: http://www.datavisguide.com.

Table 1. Results of the competitiveness test, our model is labeled NEViM [2].

Data set	Description	Records	Question	Used attributes	Excel	Google sheets	Tableau	Voyager	Watson analytics	NEViM
1	Favourite subjects within a class of students	7	What does the composition of the data look like?	Subject, no. of students	Bar chart, pie chart	Bar chart, pie chart	Bar chart	Bar chart	Bar chart	Bar chart
2	Average prices of cigarettes over several years	8	What was the development of the cigarette price over the years?	Year, average price	line chart, bar chart	Line chart	Line chart	Bar chart	None	Line chart
3	Percentage of men and women in EU countries for 2016	28	Which 5 countries have the highest percentage of females?	Country, % of men, % of women	Clustered bar chart, scatter plot, stacked bar chart	Clustered bar chart	Proportional symbol map	Scatter plot	Clustered bar chart	Clustered bar chart
4	Causes of death in Kenya in 2012	12	How big of a part does each cause take?	Cause of death, no. of deaths, % of total	None	Pie chart	Tree map	None	None	Tree map
5	Daily ice cream sales information with temperature	30	Are ice cream sales related to the weather?	Income, temperature	Scatter plot, clustered bar chart, line chart, stacked bar chart	Line chart, scatter plot, clustered bar chart	Scatter plot	Scatter plot	Scatter plot	Scatter plot
6	Email communication between researchers working together	461	Which researcher is connected to most people?	Sender, receiver	None	None	None	Scatter plot	None	Network
7	Finishing times of runners in the 2014 Boston Marathon	32K	Which finishing time interval was the most common?	Finishing time	Scatter plot, line chart	Line chart, histogram	Histogram	None	Histogram	Histogram
8	Records of UFO sightings with detailed information	80K	Are there any clusters of locations where UFOs have been seen more often?	Latitude, longitude	None	None	None	None	None	Dot map
9	List of cars and their parameters	393	Are there any relationships between the different parameters?	Miles per gallon, no. of cylinders, displacement, horsepower, weight, acceleration, year	Stacked line chart	None	None	None	None	Parallel coordinates
10	Origins and destinations of flights within the US	4K	Which city has the most ingoing and outgoing flights?	Flight origin, flight destination	None	None	Proportional symbol map	None	None	Connection map

such solutions from our test results because we feel that it is a workaround. For Microsoft Excel and Google Sheets, the recommendation process results in several recommendations and the systems do not rank them. For these cases we recorded all valid recommendations.

Results. For data set 1, all systems recommended a bar chart. Excel and Google Sheets also recommended a pie chart. The recommendations for data set 2 were either line charts or bar charts. The specified question could be answered by either of these. Watson Analytics was not able to give a recommendation because it could not recognize that the average price attribute was a number. We have attempted resolving this issue but were not able to. For data set 3, the majority recommendation was a clustered bar chart, in line with the recommendation made by NEViM. Data set 4 proved to be challenging for Voyager and Watson Analytics. Since the data was hierarchical and the question was asking to see parts-of-whole, a suitable solution would be a tree map. A pie chart shows parts-of-whole, but does not indicate hierarchy. The question asked for data set 5 could be answered using different types of data visualizations. Since it is asking to analyze the correlation between 2 variables, a scatter plot is a suitable solution. All systems recommended it. Data set number 6 was an example of a social network, thus the most suitable visualization would be a network. However, the answer to the specified question could also be answered with a scatter plot as suggested by Voyager. This is because networks can also be represented as adjacency matrices and the scatter plot generated by Voyager is essentially an adjacency matrix. Data set 7 and its question were aimed at visualizing distributions. Distributions can be visualized, among others with histograms, scatter plots and line charts. Data set 8 was an example of spatial data. Spatial data is best visualized through maps. Tableau offers map visualizations but we suspect that it cannot plot on the map according to latitude and longitude coordinates. Watson Analytics and Google Sheets have the same issue. Microsoft Excel and Voyager do not support maps at all. In Data set 9 the answer to the question was revealed through comparing 7 attributes. This meant that the visualization has to support 7 different variables. Both stacked line chart and parallel coordinates are valid solutions. The final data set 10 was again spatial. This time it could be solved through plotting on a map but also by analyzing the distribution of the data set. Both proportional symbol map and connection map (as a flight implies a connection between two cities) are valid solutions.

8.2   Extendibility Test

The aim of this test was to determine whether our model was easily extendible. We went through the process of adding a new data visualization type - a Sankey diagram. Sankey diagrams are flow diagrams that display quantities in proportion to one another. An example of a Sankey diagram can be seen in Fig. 2.

We look into the classifications that we already have and search for the most similar one. We find out that the Tree Map has the same classification, so we

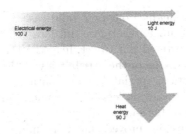

Fig. 2. Example of a Sankey diagram showing the distribution of energy in a filament lamp [30].

need to find a distinguishing feature between a Tree Map and a Sankey diagram. That feature is, that a Sankey diagram shows flow. We search through the model and find occurrences of a Tree Map. We then add a question asking "Do you want to show flow?". If the user answers "Yes", he gets a recommendation for a Sankey diagram. If he answers "No" he gets a Tree Map.

9   Evaluating the Model

9.1   Test Results

For the competitivness test, we can observe that NEViM provided usable solutions in all cases. The users have several paths that they can take through NEViM to get to a recommendation, depending on what information they know about their data or their task. NEViM has an advantage that it is not limited by implementation. Since two of our data sets were aimed at spatial data visualization and one at network data visualization, some systems were not able to make recommendations simply because they do not support such visualization types. Furthermore, NEViM includes more types of visualizations than any of the current systems, which results in recommendations for specialty visualizations that can be more suitable for a certain task. Another advantage is that it always results in only one recommendation, unlike Microsoft Excel or Google Sheets, where the user has to choose which one out of the set of recommendations to use. According to our survey, the most used visualization tool which incorporates a recommender system is Tableau (28% of non-expert respondents). From the

Fig. 3. Two possible paths to reach a Sankey diagram (left: task-based, right: data-based) [2].

result table, we can see that in 5 out of 7 valid cases, NEViM made the same recommendation as Tableau. Furthermore, in data set 3 Tableau also made a recommendation for a Clustered Bar Chart, like NEViM did, but it was not the resulting recommendation. One of the attributes was the name of a country, so Tableau evaluated the data as spatial. We have noticed that whenever there is a geographical attribute, Tableau prefers to recommend maps, even though they might not be the most suitable solution (Fig. 3).

In the extendibility test we proved that it is indeed possible to add a new data visualization into the model. The process turned out to be easy, although we must acknowledge, that the finding of a distinguishing feature might be more complex in some cases. Before adding a visualization, one must be sure that they are adding a completely new type and not just a variation of an already existing visualization. This could potentially be very hard to determine, even for experts. We believe that the classification system we have adopted has potential to help with this. New distinguishing features could be added to the classification hierarchy, for example incorporating perceptual qualities of visualizations, their suitability for a specific field (such as finance) etc. You can read more about our ideas in Sect. 12.

9.2   Fulfilling Requirements

1. **Simplicity** - Thanks to its question-based structure, using the model is simple. The user only has to answer yes or no questions. The basic structure is very straightforward.
2. **Clarity** - The result of our recommendation system is a single data visualization, making it very clear. We believe that non-expert users need a clear answer to their visualization problem. If they are given a choice between two or more visualizations in the end, we believe that we have failed at the task of recommending them the most suitable one. We have narrowed their choices, but still have not provided a clear answer. However, this decision seems to be a controversial one, so it definitely needs to be validated through a user study. In the case that none of the data visualizations within the model are determined as suitable, the model still makes a recommendation to visualize using a table.
3. **Versatility** - NEViM combines two different types of data visualization recommendation systems as defined in [8]: task-oriented and data characteristics-oriented. These two types are distinguished by two different starting points within our model. Thanks to its base structure the model can be easily implemented in various different programming languages and environments.
4. **Extendibility** - To illustrate the extendibility of the model, we have added the Sankey diagram visualization. This proved to be a doable task.
5. **Education** - This requirement has not been met yet. We feel that the fulfillment is more related to the implementation phase than the conceptual phase where we find ourselves now.
6. **Transparency** - The traversal through our model is logical enough that it is clear why a certain type of data visualization was recommended.

7. **Self-learning** - Our model is machine learning friendly and techniques can be applied for it to be able to self-learn.
8. **Competitiveness** - Through testing we have proved that our model produces recommendations similar or identical to existing solutions. It provided suitable solutions for all cases tested, unlike existing solutions.

9.3   Advantages and Disadvantages

The advantages already been discussed in Sect. 9.1 so we will not repeat them. A possible disadvantage of NEViM could be that the user has to either know what their main task is, or know what type of data they have. The question is, whether non-expert users will be able to determine this. We believe that this could be fixed through user testing to validate the overall structure of the model as well as the quality of the questions. The questions could be checked by a linguistics expert to see whether the wording is suitable and does not lead to possible ambiguous interpretations.

Another disadvantage might lie in the fact that since we use data science terminology in our questions, we risk that non-experts might not be familiar with it and might not be able to answer the question. A solution could be to clarify the terms using a dictionary definition, which could pop up when the user hovers over the unfamiliar term. The solution is more part of the implementation phase, not the theoretical phase which we discuss here.

We have questioned whether the choice to recommend a table when no other suitable visualization is found is the correct one. There is an ongoing debate about when it is best to not visualize things, as discussed by Stephanie Evergreen [31]. Within the implementation phase, data could be collected to find out in how many cases the Table option is reached, to identify whether it is necessary to further address this issue.

10   Implementing the Model

Earlier in this paper we have set a requirement for our model which stated that the model has to be easily implemented in different programming languages and environments. We will now implement the model we have constructed in a way that would allow it to be tested with users and adjusted. Since we are not very skilled in implementing a decision tree, we opted to build a prototype in a prototyping tool instead.

Once we found a suitable prototyping tool for our task. We set out to create an application, where each screen would be a question from our model. The user would then select either 'yes' or 'no' and be taken to the next question according to our model. This was an easy task, but very time consuming. Our final model had 107 internal nodes and we had to create a screen for each one. Luckily, we were able to reuse some of the screens as some of the questions occur in our model multiple times (for example once in the task-based branch and once in the data-based branch). When we finished this process, we realised another issue that our model has. The traversal through it could get very lengthy. The longest route from the root node to a leaf node was 12 questions. We came up with a way to solve this problem. We introduced the possibility of multiple choice answers rather than just 'yes' or 'no'. In Fig. 4 you can see an example of a problematic screen from the prototype. In the first version of our prototype, we would have to go through the displayed options one by one, asking the user a question about each option. By introducing the solution displayed, we were able to shorten the lengthy routes and also remove possible bias of the user stemming from the order in which the questions were asked in the initial version. This solution also has another advantage in that the user can see a summary of the options in advance and it can help him in deciding what his main task is. Lastly, during this phase we also realised that we must provide the user with a fallback solution. What if the user simply does not know what their main task is? Which option would he select? We need to account for such an option in our model.

Fig. 4. Example of a problematic screen from the prototype.

11   Discussion and Conclusions

In this study we have managed to define, specify and go through the process of building a data visualization recommender system. We evaluated our app-roach by constructing our very own data visualization recommender system from scratch. We studied existing solutions thoroughly, we conducted a survey, we put together requirements for a model, we constructed a model, we tested the model, evaluated it and then made an example implementation. Initially we aimed for our model to be suited for non-expert users, but we were surprised to see that our survey showed that there is little to no difference in the attitude of non-experts and other users towards data visualization recommender systems. Throughout the paper we still comment on ways to make the model more suitable for non-experts, but because of the results of the survey, it became a secondary focus for us.

We have proven that there is definitely a place for data visualization recom-mender systems in the data science world. We are very pleased with the model that we have constructed during the study and also with the fact that this paper can serve as a step-by-step guide for someone who wants to make a data visualization recommender system.

12   Future Work

In the future, we would like to make use of the model that we have constructed as part of this study and further expand and improve it. We would like to perform more tests with more data sets and also carry out a usability test with different types of users. An idea could be to implement the model as a web application where users could rate the resulting recommendations, suggest new paths through the model or request new visualization types to be included. This would also validate the question paths that we have designed. The final recommendation could be enhanced with useful information about the data visualization type, tips on how to construct it, which tools to use and examples of already made instances. This would transform the model into a very useful educative tool.

Another possible extension to the model could be to add another view which would incorporate information about the domain that the user's data comes from. There are data visualizations that are more suited for a specific data domain than others. For example, the area of economics has special types of data visualizations that are more suited to exposing different economic indicators. This would make the model part of the domain knowledge oriented data visualization systems recommender systems category according to [8].

We could introduce different features that could influence the visualization ranking - e.g. perceptual qualities of different data visualization types. Now that we have established a successful base, the possibilities for further development are endless.

Acknowledgements. Research supported by SGS grant No. SGS17/210/OHK3/3T/ 18 and GACR grant No. GA18-18080S.

References

1. Data-Driven Documents (d3.js). https://github.com/d3/d3/wiki/Gallery. Accessed 4 Aug 2017
2. Kubernátová, P., Friedjungová, M., van Duijn, M.: Knowledge at first glance: a model for a data visualization recommender system suited for non-expert users. In: Proceedings of the 7th International Conference on Data Science, Technology and Applications - Volume 1: DATA, INSTICC, SciTePress, pp. 208–219 (2018)
3. O'Neil, C., Schutt, R.: Doing Data Science: Straight Talk From The Frontline. O' Reilly Media, Sebastopol (2014)
4. Tukey, J.W.: Exploratory Data Analysis. Addison-Wesley, Reading (1970)
5. Kirk, A.: Data Visualization: A Handbook for Data Driven Design. SAGE, London (2016)

6. Illinsky, N., Steele, J.: Designing Data Visualizations: Representing Informational Relationships. O'Reilly Media, Sebastopol (2011)
7. Munzner, T., Maguire, E.: Visualization Analysis and Design. CRC Press, Boca Raton (2015)
8. Kaur, P., Owonibi, M.: A review on visualization recommendation strategies. In: Proceedings of the 12th International Joint Conference on Computer Vision, Imaging and Computer Graphics Theory and Applications, pp. 266–273 (2017)
9. Gnanamgari, S.: Information presentation through default displays. Ph.D. Dissertation, Philadelphia (1981)
10. Mackinlay, J.: Automating the design of graphical presentations of relational information. ACM Trans. Graph. (TOG) **5**(2), 110–141 (1986)
11. Casner, S., Larkin, J.H.: Cognitive efficiency considerations for good graphic design. Carnegie-Mellon University Artificial Intelligence and Psychology Project, Pittsburgh (1989)
12. Roth, S.F., Mattis, J.: Data characterization for intelligent graphics presentation. In: SIGCHI Conference on Human Factors in Computing Systems (1990)
13. Hanrahan, P.: VizQL: a language for query, analysis and visualization. In: Proceedings of the 2006 ACM SIGMOD international conference on Management of data. ACM (2006)
14. Stolte, C.: Polaris: A system for query, analysis, and visualization of multidimensional relational databases. IEEE Trans. Vis. Comput. Graph. **8**(1), 52–65 (2002)
15. Mackinlay, J., Hanrahan, P., Stolte, C.: Show me: automatic presentation for visual analysis. IEEE Trans. Vis. Comput. Graph. **13**(6), 1137–1144 (2007)
16. Viegas, F., Wattenberg, M., van Ham, F., Kriss, J., McKeon, M.: ManyEyes: a site for visualization at internet scale. IEEE Trans. Vis. Comput. Graph. **13**(6), 1137–1144 (2007)
17. Smart data analysis and visualization. https://www.ibm.com/watson-analytics. Accessed 4 Aug 2017
18. Key, A., Perry, D., Howe, B., Aragon, C.: VizDeck: self-organizing dashboards for visual analytics. In: Proceedings of the 2012 ACM SIGMOD International Conference on Management of Data (2012)
19. Available chart types in Office. https://support.office.com/. Accessed 4 Aug 2017
20. Vartak, M., Madden, S., Parameswaran, A., Polyzotis, N.: SeeDB: supporting visual analytics with data-driven recommendations. VLDB **8**, 2182–2193 (2015)
21. Wongsuphasawat, K., Moritz, D., Mackinlay, J., Howe, B., Heer, J.: Voyager: exploratory analysis via faceted browsing of visualization recommendations. IEEE Trans. Vis. Comput. Graph. **22**(1), 649–658 (2016)
22. Vega Compass. https://github.com/vega/compass. Accessed 4 Aug 2017
23. Satyanarayan, A., Moritz, D., Wongsuphasawat, K., Heer, J.: Vega-Lite: a grammar of interactive graphics. IEEE Trans. Vis. Comput. Graph. **23**(1), 341–350 (2017)
24. Chart and Graph Types. https://support.google.com/. Accessed 9 Aug 2017
25. Wehrend, S., Lewis, C.: A problem-oriented classification of visualization techniques. In: Proceedings of the 1st Conference on Visualization 1990. IEEE Computer Society Press (1990)
26. Zhou, M.X., Feiner, S.K.: Visual task characterization for automated visual discourse synthesis. In: Proceedings of the SIGCHI conference on Human factors in computing systems. ACM Press/Addison-Wesley Publishing Co. (1998)
27. Gotz, D., Wen, Z.: Behavior-driven visualization recommendation. In: Proceedings of the 14th International Conference on Intelligent User Interfaces. ACM (2009)
28. Zhou, M.X., Chen, M., Feng, Y.: Building a visual database for example-based graphics generation. In: INFOVIS 2002 IEEE Symposium (2002)

29. Alborzi, F., Reutter, J., Chaudhuri, S.: DataSlicer: task-based data selection for visual data exploration. arXiv preprint (2017)
30. Bbccouk: http://www.bbc.co.uk/schools/gcsebitesize/science/aqa/energy efficiency. Accessed 17 Aug 2017
31. Evergreen, S.D.: Effective Data Visualization: The Right Chart for Your Data. SAGE Publications, Thousand Oaks (2016)
32. Yau, N.: Visualize This: The FlowingData Guide to Design, Visualization, and Statistics. Wiley, Hoboken (2011)
33. Yau, N.: Data Points: Visualization That Means Something. Wiley, Hoboken (2013)
34. Heer, J., Bostock, M., Ogievetsky, V.: A tour through the visualization ZOO. Queue 8(5) (2010)
35. Hardin, M., Hom, D., Perez, R., Williams, L.: Which chart or graph is right for you?. Tell Impactful Stories with Data, Tableau Software (2012)
36. Yuk, M., Diamond, S.: Data Visualization for Dummies. Wiley, Hoboken (2014)
37. Brath, R., Jonker, D.: Graph Analysis and Visualization: Discovering Business Opportunity in Linked Data. Wiley, Hoboken (2015)
38. Borner, K., Polley, D.E.: Visual Insights: A Practical Guide to Making Sense of Data. MIT Press, Cambridge (2014)
39. Telea, A.C.: Data Visualization: Principles and Practice. CRC Press, Boca Raton (2007)
40. Borner, K.: Atlas of Knowledge: Anyone Can Map. MIT Press, Cambridge (2015)
41. Ware, C.: Visual Thinking: For Design. Morgan Kaufmann, Burlington (2010)
42. Ware, C.: Information Visualization: Perception for Design. Elsevier, Amsterdam (2012)
43. Stacey, M., Salvatore, J., Jorgensen, A.: Visual Intelligence: Microsoft Tools and Techniques for Visualizing Data. Wiley, Hoboken (2015)
44. Hinderman, B.: Building Responsive Data Visualization for the Web. Wiley, Hoboken (2015)
45. Gemignani, Z., Gemignani, C., Galentino, R., Schuermann, P.: Data Fluency: Empowering Your Organization with Effective Data Communication. Wiley, Hoboken (2014)

A Comprehensive Prediction Approach for Hardware Asset Management

Alexander Wurl[1]([⊠])(iD), Andreas Falkner[1](iD), Peter Filzmoser[3](iD),
Alois Haselböck[1](iD), Alexandra Mazak[2](iD), and Simon Sperl[1](iD)

[1] Siemens AG Österreich, Corporate Technology, Vienna, Austria
{alexander.wurl,andreas.a.falkner,alois.haselboeck,
simon.sperl}@siemens.com
[2] TU Wien, Business Informatics Group, Vienna, Austria
mazak@big.tuwien.ac.at
[3] Institute of Statistics and Mathematical Methods in Economics,
TU Wien, Vienna, Austria
p.filzmoser@tuwien.ac.at

Abstract. One of the main tasks in hardware asset management is to predict types and amounts of hardware assets needed, firstly, for component renewals in installed systems due to failures and, secondly, for new components needed for future systems. For systems with a long lifetime, like railway stations or power plants, prediction periods range up to ten years and wrong asset estimations may cause serious cost issues. In this paper, we present a prediction approach combining two complementary methods: The first method is based on learning a well-fitted statistical model from installed systems to predict assets needed for planned systems. Because the resulting regression models need to be robust w.r.t. anomalous data, we analyzed the performance of two different regression algorithms – Partial Least Square Regression and Sparse Partial Robust M-Regression – in terms of interpretability and prediction accuracy. The second method combines these regression models with a stochastic model to estimate the number of asset replacements needed for existing and planned systems in the future. Both methods were validated by experiments in the domain of rail automation.

Keywords: Predictive asset management ·
Obsolescence management Robust regression · Data analytics

This work is funded by the Austrian Research Promotion Agency (FFG) under grant 852658 (CODA).
Alexandra Mazak is affiliated with the Christian Doppler Laboratory on Model-Integrated Smart Production at TU Wien.

C. Quix and J. Bernardino (Eds.): DATA 2018, CCIS 862, pp. 26–49, 2019.
https://doi.org/10.1007/978-3-030-26636-3_2

1   Introduction

A crucial task in Hardware Asset Management[1] is the prediction of (i) the numbers of assets of various types for planned systems, and (ii) the numbers of assets in installed systems needed for replacement, either due to end of lifetime (preventive maintenance) or due to failure (corrective maintenance). This is especially important for companies that develop, engineer, and sell industrial and infrastructural systems with a long lifetime (e.g., in the range of decades) like power plants, factory equipment, or railway interlocking and safety systems.

Predicting the number of assets for future projects is important for sales departments to estimate the potential income for the next years and for bid groups to estimate the expected system costs. To predict assets, simple linear regression models achieve only sub-optimal results when heterogeneous data sources, faulty data warehouse entries, or non-standard conditions in installed systems are involved [9]. Therefore, we evaluate regression methods which are robust against anomalies introduced by non-standard conditions and faulty data so that the learned model can provide optimal prediction results for future projects.

Service contracts oblige a vendor to guarantee the functioning of the system for a given time period with a failure rate or system down-time lower than a specified value. This implies that all failing hardware assets (and often also those expected to fail in near future) must be replaced without delay in order to ensure continuous service. A precise prediction of the number of asset types needed in the next months or years is essential for the company to calculate service prices and to fulfill the service obligations for replacing faulty components. Often, such predictions are made by experts relying on their experiences and instinct, including a certain safety margin. A more informed prediction is based on expectation values: Add the number of assets in existing systems and the expectation values for the numbers of assets for planned systems to get a basis for calculating the expectation values of asset replacements. This kind of prediction is simple, but it cannot provide information about the confidence interval of the prediction. A prediction far below the actual need may cause that assets are not available in time with all the annoying consequences for the vendor such as penitent fees and a bad reputation. A prediction far beyond the actual need will bind costs in unnecessary asset stocks. We provide a predictive model for the asset replacements, taking into account all necessary data from installed systems, project predictions and renewal necessities. The result is a probability function representing not only the predicted number of assets but also the uncertainty of the prediction.

Figure 1 sketches our framework for combining both kinds of predictions. This paper is an extended version of [33] which was presented at the DATA 2018 conference. In the first step, we learn a regression model, which is capable to predict the number of assets for planned systems based on data about systems

---

[1] Please note, that in this article we use the term "asset" in the sense of physical components such as hardware modules or computers but not in the sense of financial instruments.

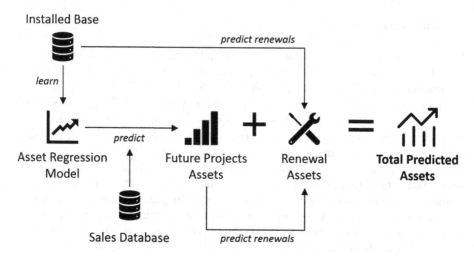

Fig. 1. Asset prediction framework: combining asset predictions for planned systems with prediction of asset renewals for all systems.

in the installed base. Analyzing the performance of the regression model, in this extended work we integrate a robust method and evaluate results according to prediction accuracy. The second step provides a predictive model for the total number of assets needed in the next n years, comprising new assets (for planned systems) and asset replacements (for installed and planned systems). We contribute extended work by adequately presenting results for daily business, i.e., an advanced visualization of prediction results supports technical engineers to quickly grasp all important information.

This paper is organized as follows: Sect. 2 presents the *asset regression model* for predicting the assets of a planned system. Section 3 presents the *asset prediction model* for estimating the total number of assets (new and replacements) needed in the next n years. Section 4 presents related work and Sect. 5 concludes the paper.

2 Asset Regression Model

In this section, we present a strategic learning method for predicting the number of assets of a given type needed for a planned system. To achieve this, an asset regression model is learned from the relation between feature and asset numbers in installed systems (in previous projects). Features are properties of a system or project that can be counted or measured by domain experts. For instance, in the domain of rail automation, features are the numbers of track switches, of signals of various types, of railway crossings, etc. In large systems, there are hundreds of different features and asset types, whereas the number of installed systems may be lower than hundred. In order to increase the prediction accuracy of assets needed in future projects, we need to find those features that still can predict the number of assets of a given type with sufficient accuracy.

In a first step, the goal is to model the relationship between features and one selected asset. In other words, let

$$
D = \begin{pmatrix}
x_{11} & \cdots & x_{1p} & y_{11} & \cdots & y_{1d} \\
x_{21} & \cdots & x_{2p} & y_{21} & \cdots & y_{2d} \\
\vdots & \ddots & \vdots & \vdots & \ddots & \vdots \\
x_{n1} & \cdots & x_{np} & y_{n1} & \cdots & y_{nd}
\end{pmatrix} = (XY) \tag{1}
$$

be an arbitrary data set depicted as data matrix. Each observation with index i ($1 \leq i \leq n$), stored in the i-th row in D, corresponds to an existing system and consists of input variable values $x_i = (x_{i1}, \ldots, x_{ip})^T$ (features) and related output variable values $y_{i1}, \ldots, y_{id}$ (assets). Assume that the inputs are collected in the data matrix $X \in \mathbb{R}^{n \times p}$ with entries x_{ij}, and the outcomes in the matrix $Y \in \mathbb{R}^{n \times d}$, with entries y_{ik}. Further, we assume that the columns of X are mean-centered.

In the following we will focus on predicting a specific output variable y_k, for $k \in \{1, \ldots, d\}$, which is the k-th column of Y. In order to simplify the notation, we will denote $y := y_k$ in the following, referring to a model for a univariate outcome variable.

In the task of asset estimation we assume a linear relationship between the predictors X (features) and the predictand y (assets) related to (1),

$$
y = X\beta + \epsilon \tag{2}
$$

where $\beta = (\beta_1, \ldots, \beta_p)^T$ is the vector of regression coefficients, and $\epsilon = (\epsilon_1, \ldots, \epsilon_n)^T$ is the error term.

Although in a preprocessing step for integrating heterogeneous data into a data warehouse [31,32], a high degree of data quality is achieved, the data set may contain outliers. Outliers could be atypical values in the considered asset, or they could relate to atypical feature values which exhibit deviations far from a linear relationship [7]. They can occur during data fusion of heterogeneous data sets, by manual changes in data sets, or in non-standard system variants. In a first attempt, we apply the method of Partial Least Squares Regression (PLSR) [30], which is still not robust against these types of outliers. Later, more advanced techniques are applied that are able to automatically downweight outliers.

2.1 Partial Least Squares Regression (PLSR)

PLSR is a regression method that reduces the input variables (i.e., the features) to a smaller set of uncorrelated components and performs least squares regression on these components. This technique is especially useful when features are highly collinear, or when the data set reveals more features than observations (i.e., installed systems in previous projects). Therefore, this is a promising technique for our setting, where the number of input variables is large compared to the number of observations. It is known that in such a situation, ordinary least squares regression leads to models with poor prediction performance [30].

For PLSR we assume the regression model from Eq. (2), which can be written in terms of the i-th observation as

$$y_i = x_i^T \beta + \epsilon_i, \tag{3}$$

for $i = 1, \ldots, n$. Instead of directly estimating the p-dimensional vector β of regression coefficients, PLSR assumes a so-called latent variable model

$$y_i = t_i^T \gamma + \epsilon_i^*, \tag{4}$$

with q-dimensional score vectors t_i and regression coefficients γ, and an error term ϵ_i^*. The dimension q is smaller than (or equal to) p, typically even much smaller, and thus estimating the coefficients γ in (4) will be more stable. The scores t_i in (4) are defined intrinsically through the construction of latent variables. This is done sequentially, for $k = 1, 2, \ldots, q$, by using the criterion

$$a_k = \underset{a}{\mathrm{argmax}}\, Cov(y, Xa) \tag{5}$$

under the constraints $\|a_k\| = 1$ and $Cov(Xa_k, Xa_j) = 0$ for $1 \le j < k$. The vectors a_k with $k = 1, 2, \ldots, q$ are also called *loadings*, which are finally collected in the columns of the matrix A. The resulting score matrix is then

$$T = XA, \tag{6}$$

and it contains the score vectors t_i from (4) in the rows. Since $t_i = x_i^T A$, for $i = 1, \ldots, n$, it follows from (4) and (3) that $\beta = A\gamma$, and thus one obtains the regression coefficients in the original model.

2.2   Iterative Learning Using PLSR

One of the major challenges in our problem domain comes from potential outliers of asset quantities in the data set. Since the input data originate from heterogeneous data sources, anomalies may occur due to the complexity of merging data structures and due to non-standard conditions in some projects. Even if high data quality during the process of data integration is achieved by approaches such as described in [32], anomaly detection including model validation is required during the training phase of the prediction model.

In model validation, the most common methods for training regression models are Cross-Validation (CV) and Bootstrapping. In our setting, CV is preferred because it tends to be less biased than Bootstrapping when selecting the model and it provides a realistic measurement of the prediction accuracy.

To ensure a stable prediction model, we make use of a double cross-validation strategy which is a process of two nested cross-validation loops. The inner loop is responsible for validating a stable prediction model, the outer loop measures the performance of prediction. This strategy is applied to similar approaches that have been described for optimizing the complexity of regression models in chemometrics [4], for a binary classification problem in proteomics [23], and for a discrimination of human sweat samples [3]. Our approach is a formal, partly new combination of known procedures and methods, and has been implemented in a function for the programming environment R[2]: In an internal cross-validation loop, we train our model with 80% of the data, using 10-fold cross-validation. In an external cross-validation, we test our trained model with the 20% rest of the full data set.

Training. Our regression model potentially uses all input variables of all observations from the training set to predict the selected output variable y. Call I_{train} the index set containing all indexes of the training set observations, and n_{train} the number of training set observations. Following PLS regression, the estimated parameters $\hat{\gamma}$ are obtained via (4) using $q \in \{1, \ldots, p\}$ latent variables, leading to the estimated parameters $\hat{\beta}$ in (3). We denote the corresponding predicted values by $\hat{y}_i^{(q)}$, for $i \in I_{train}$, which should be as close as possible to the real asset values y_i. "As close as possible" is achieved by an iterative process of improving the prediction model step by step, which is computed by means of PLSR in combination with CV.

In the inner CV loop, we employ 10-fold CV to train the model. By training, we aim at finding the number q of components which adequately explains both predictors and response variances. This is calculated by the Root Mean Squared Error (RMSE) for a model with q components ($1 \leq q \leq p$):

$$RMSE_{CV}(q) = \sqrt{\frac{1}{n_{train}} \sum_{i \in I_{train}} \underbrace{(y_i - \hat{y}_i^{(q)})^2}_{\text{quadratic error}}}$$

Aiming for an optimal number of components, the goal is to find the number of components $q = c_{opt}$ with the lowest cross-validated RMSE. The minimum indicates the number of components leading also to a minimal prediction error for new test set observations.

In the outer loop, in addition to $RMSE_{CV}$ we use the multiple R-square measure R^2 with the optimal number c_{opt} of PLSR components. Denote by I_{test} the index set with the test set observations kept for the outer loop. Then the R^2 is defined as

$$R^2 = 1 - \frac{\sum_{i \in I_{test}} (y_i - \hat{y}_i^{(c_{opt})})^2}{\sum_{i \in I_{test}} (y_i - \overline{y})^2}, \tag{7}$$

with the arithmetic mean $\overline{y}$ of the test set observations of the response variable. The R^2 is a value in the interval $[0, 1]$, and it measures how much variance of the response is explained by the predictor variables in the regression model.

While calculating $RMSEP_{CV}(q)$, the estimated CV error usually decreases with an increasing number of components, and increases once q is getting closer to p. In case of an error increase for small q, it is likely that:

1. Dependent and independent variables are not linearly related enough to each other.

---

[2] www.r-project.org.

2. The data set contains insufficient number of observations to reveal the relationship between input and output variables.
3. There is only one component needed to model the data.

On the other hand, a low R^2 value indicates that there are still outliers in the observations we used for training the model (or in the test set observations). These outliers in the data degrade the quality of our model, so we remove them and train a new model without them.

Algorithm 1 sketches the iterative process of training a model for asset prediction [33]. Firstly, the data set is shuffled to obtain a fair distribution of observations. Secondly, a PLSR model is trained and validated through RMSE, using $1 \leq q \leq p$ components, and the model with the optimal number of components c_{opt} is returned. Thirdly, this model is applied to the test data set and R^2 is computed. We use two criteria to decide whether our model is already good enough or not: (i) the RMSE must monotonically decrease for $1 \leq q \leq c_{opt}$ components, and (ii) according to longitudinal studies R^2 must be at least 0.9 [15]. If the model is not yet good enough, we remove all observations from the data set that we categorized as outliers. After some experiments, the following outlier classification has turned out to be useful: Outliers are observations with total residuals that are larger than 25% of the maximum asset value in the data set. Accordingly, we eliminate observations with index i if,

$$|\hat{y}_i - y_i| > 0.25 * y_{max} \tag{8}$$

After removing the noisy observations, learning starts again. We proceed with this learning cycle until our model meets the above mentioned quality criteria or no improvement could be achieved.

In the last step of Algorithm 1, the features that "significantly" contribute to a prediction are extracted according to their weights. Currently, this selection is done by a domain expert based on the weights of the features that dominate the prediction model.

Our trained model is now

$$\hat{y}_a = f_a(X') \tag{9}$$

where $X' \subseteq \{x_1, \ldots, x_d\}$ is the set of significant features used for predicting assets of type a based on a model f_a. Applying this model to concrete feature values $\bar{X}'$ of features X' for a planned system will provide an expectation value for the number of assets a. In this sense, the output variable $\hat{y}_a$ could be seen as a random variable with a normal distribution - $\hat{y}_a \sim \mathcal{N}(\mu, \sigma^2)$ - where the mean value μ is $f_a(\bar{X}')$ and σ is the above described root mean squared error RMSEP. In the next subsection, we demonstrate this on an example.

2.3   Example and Experimental Results

We tested the validity of our method on a data set from the railway safety domain with data collected over about a decade. Our data set contained ca.

Algorithm 1. Calculate asset regression model. (Source: [33]).

---

1. Shuffle rows in the data set to obtain a fair distribution of observations.
2. Train a PSLR model with $1 \leq q \leq p$ components for the training data
   (a) Use 10-fold cross-validation and compute the resulting errors $RMSE_{CV}(q)$.
   (b) Determine the numbers of components c_{opt} for which $RMSE_{CV}(q)$ is minimal, and return this model.
3. Test the model on test data set by computing R^2.
   If RMSEP does not monotonically decrease with the number of components,
   or if $R^2 < 0.9$
   (a) Identify outliers in observations of training and test data according to Eq. (8) and remove them.
   (b) Go to step 2.
4. Extract "significant" features to select the final prediction model.

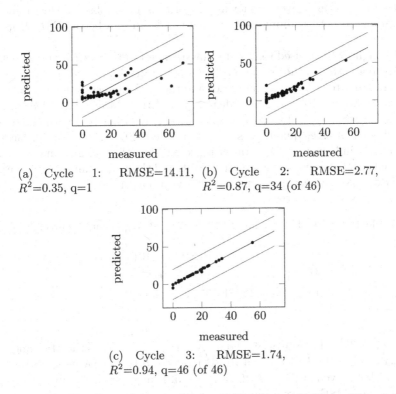

(a) Cycle 1: RMSE=14.11, R^2=0.35, q=1

(b) Cycle 2: RMSE=2.77, R^2=0.87, q=34 (of 46)

(c) Cycle 3: RMSE=1.74, R^2=0.94, q=46 (of 46)

Fig. 2. Three cycles for training a regression model for asset $A41$. The red lines indicate the outlier rule from Eq. (8). (Source: [33]).

140 features (input variables), ca. 300 assets (potential output variables), and ca. 70 observations (installed systems). We chose the concrete asset type $A41$ - a hardware module that controls track switches - to demonstrate training of the regression model and prediction. We implemented this example and the regression training algorithm in R, making use of the plsr() function with the NIPALS algorithm [28, 29] as implemented in the package pls[3].

Before starting any calculation, we preprocessed the data set by removing all assets except $A41$ and shuffled the data set in order to have a fair distribution of the training data set and the test data set. Following Algorithm 1, we built the model by specifying $A41$ as the output variable and all features as input variables.

Figure 2 shows the three training cycles for $A41$, each of them contains all data points. After the first cycle, four observations were identified as outliers and removed from the whole data set. The RMSE is relatively high with only one component q needed. After the second cycle, another two observations were removed. The RMSE decreases with an optimal amount of 34 out of 46 available components q. After the third cycle we ended up in a regression model of high precision. The RMSE further decreases as the number of optimal components q raises to 46. Conclusively, following Algorithm 1, we observe that removing observations in both training and test data ends up in a better $RMSE_{CV}$ and R^2.

Although PLSR is used to identify the strongest outliers, we assume that our data set may still contain some outliers after the training and outlier removal cycles. In order to indicate the behavior and the performance of models including outliers, we use the 10%-trimmed Mean Square Error (MSE). The 10%-trimmed MSE is the MSE applied to 90% of the data set, ignoring 10% with the highest squared error of asset quantity estimations. These 10% of data would have the biggest error influence. If the data set is skewed then the trimmed mean is closer to the bulk of the observations [27]. The MSE and the 10%-trimmed MSE applied to our example data set including all data with outliers are shown in Table 1.

Table 1. MSE and 10%-trimmed MSE of PLSR regression. (Source: [33]).

	PLSR	PLSR (3 cycles)
MSE	84.05	105.20
10%-trimmed MSE	35.67	24.09

Discussion. Based on the original data set, Table 1 compares the MSE and 10%-trimmed MSE of both the initial PLSR model and the PLSR model after three cycles with removed outliers. Although the MSE increases after three cycles, the 10%-trimmed MSE shows a much better performance. This indicates that after three cycles outliers have been identified. The model learned with Algorithm 1

---

[3] http://mevik.net/work/software/pls.html.

shows that the assumption that 10% of the data set includes most of the outliers holds. The approach, however, can also become unreliable, because outliers could already influence the estimation of the regression parameters in the PLSR model, and they could be masked in a subsequent diagnostic [14]. Furthermore, removing too many observations (i.e. in several cycles) is problematic when the data set gets too small and is no longer representative.

Another challenge in leveraging prediction accuracy is the feature selection process, i.e., analyzing the variable importance. Since the PLSR model includes all coefficients, i.e., all features, mostly there exist cases without a significant difference between feature values, thus making it difficult to identify informative features. Conclusively, this naturally triggers the question when and under which circumstances a feature is informative.

Continuing the strategy of finding a latent variable model, in the following we present a robust method which is capable to provide both a robust model in the presence of outliers and an amount of informative features contributing to the prediction accuracy of an asset.

2.4 Sparse Partial Robust M-Regression

Generally, when a model for linear regression according to Eq. (2) is considered, the responses predicted by PLSR are $\hat{y} = X\hat{\beta}$. When p is large, X may contain columns of uninformative variables, i.e., variables that hold no relevant information related to the predictand. Consequently, there exists a subset of coefficients $\{\hat{\beta}_{j_1}, \ldots, \hat{\beta}_{j_{\tilde{p}}}\}$ which are small but not exactly zero, and each of them still contributes to the model and, more importantly, to increased prediction uncertainty. Instead of cutting out $\tilde{p}$ uninformative coefficients like in the previous method, a sparse estimator of β will have many coefficients that are exactly equal to zero. In order to include robustness towards outliers, we apply Sparse Partial Robust M-regression (SPRM), which provides estimates with a partial least squares alike interpretability that are sparse and robust with respect to both vertical outliers (outliers in the response) and leverage points (outliers in the space of the predictors) [7].

The SPRM estimator is obtained as follows (for details we refer to [7]):

1. Case weights $w_i \in [0, 1]$, for $i = 1, \ldots, n$, are assigned to the rows of X and y. If an observation has a large residual, or is an outlier in the covariate in the latent regression model, this observations will receive a small weight. The case weights are initialized at the beginning of the algorithm.
2. The weights from the previous step are incorporated in the maximization of (5), by weighting the observations, and thus maximizing a weighted covariance. In addition to that, an L_1 penalty is employed, which imposes sparsity in the resulting vectors a_k, for $k = 1, \ldots, q$. The result is a sparse matrix of robustly estimated direction vectors A, and scores $T = XA$.

3. The regression model (4) is considered, but the regression parameters are estimated by robust M-regression,

$$\hat{\gamma} = \underset{\gamma}{\text{argmin}} \sum_{i=1}^{n} \rho(y_i - t_i^T \gamma). \tag{10}$$

The function ρ is chosen to reduce the influence of big (absolute) residuals $y_i - t_i^T \gamma$, see [22]. Note that the least squares estimator would result with a choice $\rho(u) = u^2$, with an unbounded influence of big values u^2. The updated weights are based on $w(u) = \rho'(u)/u$, where ρ' is the derivative of the function ρ.

Steps 2 and 3 are iterated until the estimated regression coefficients stabilize. Note that there are now two tuning parameters: the number q of components, and the sparsity parameter, later on called η ("eta"); η needs to be selected in $[0, 1]$, where $\eta = 0$ leads to a non-sparse solution, and bigger values of η to more and more sparsity.

To find a suitable sparse partial robust M-regression model, we perform 10-fold cross validation applied to the complete data set for the selection of the number of components and the sparsity parameter. In the evaluation process of SPRM[4], Fig. 3 provides a graphical overview. The colors in the upper diagram represent the values of the 10%-trimmed Mean Squared Error of Prediction: the darker the blue, the lower the prediction error. We observe that in a given range of 10 components, the minimum is achieved with 10 PLS components and for a value of "etas" equal to 0. This choice of "eta" would not yield any sparsity, and thus we decided to take 8 components and "eta" equal to 0.8, since the MSEP for this combination is almost unchanged. The lower diagram shows the number of non-zero coefficients for different choices of "eta": the darker the green, the higher the eta, i.e. the more features will be set to zero. Both diagrams together show the consequences of the settings used in the SPRM regression model.

Figure 4 shows the comparison between the learned PLS model after 3 cycles in Fig. 4(a) and the learned SPRM model in Fig. 4(b) applied to the original data set. Comparing both models, a few differences can be observed. In the PLSR model the majority of predicted values is gathering close to the measured values but still with a certain discrepancy. In contrast to the SPRM model, both predicted and measured values are relatively fitted. These differences mainly show that in PLSR after 3 cycles the worst outliers have been removed, which was not necessary in SPRM to get a proper model. Furthermore, to enlighten the performance of outlier detection in SPRM, observations, which reveal a weight (w) lower than 0.5 are marked with red crosses. While lower weighted observations in SPRM are somehow excluded from higher weighted observations, a few of the same lower observations can be identified between higher weighted observations in PLSR. This indicates that despite of the iterative PLSR approach some outliers may have not been identified.

---

[4] Implemented with the sprm package in R - https://cran.r-project.org/web/packages/sprm/.

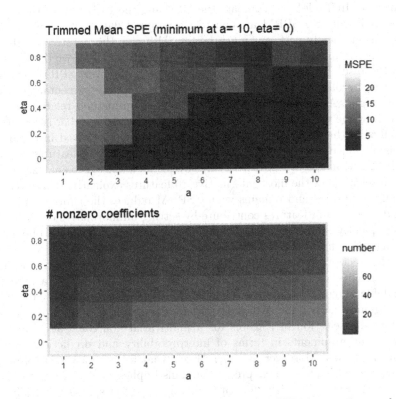

Fig. 3. A graphical orientation for finding a suitable SPRM regression model

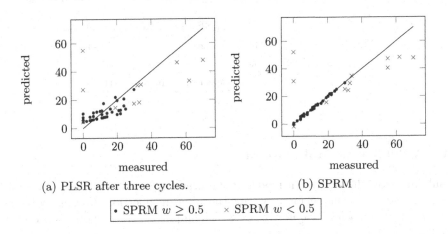

(a) PLSR after three cycles.   (b) SPRM

| • SPRM $w \geq 0.5$ × SPRM $w < 0.5$ |

Fig. 4. Comparing the models of PLSR and SPRM.

Discussion. In Table 2, we compare the MSE and the 10%-MSE of the learned models of PLSR and SPRM applied to the original data set. It appears that SPRM automatically performs much better. Although the iterative PLSR approach aims for finding a better model than just applying PLSR on data, according to the result it seems to take too much effort. Moreover, SPRM acts more objectively because cutting outliers iteratively by a certain percentage in PLSR remains vague. As we have found a proper regression method related to a univariate outcome variable, i.e. one asset, we forgo comparing other asset values.

Analyzing the quantity of coefficients (i.e., features) involved in the model, there is obviously a great difference between the quantity of features in PLSR and SPRM. Although in PLSR with three cycles we managed to analyze features within a selection of the most informative 10 features explicitly, this selection is manually performed and remains vague. SPRM reduces the effort analyzing to which degree certain features contribute by setting uninformative coefficients to 0. Consequently, sparse and robust modeling with SPRM appears to be a great advantage in terms of interpretability and prediction accuracy.

In this section we solely focused on a data set referring to a model for a univariate outcome variable. Basically, PLSR enables to analyze a model for all available outcome variables, i.e., multivariate outcome, contrarily to SPRM. At first glance, such a multivariate approach seems to be a all-in-one solution regarding the selection of features for a multivariate outcome at a time, but advantageous arguments in terms of interpretability and prediction accuracy remain unclarified. Moreover, it is very likely that for each asset, different features will be relevant for their prediction. This implies that applying SPRM to each asset individually with the appropriately adjusted sparsity will allow for more targeted interpretation, and possibly also for better prediction accuracy.

Table 2. MSE and 10%-trimmed MSE of PLSR and SPRM regression methods.

	PLSR	PLSR (3 cycles)	SPRM
MSE	84.05	105.20	71.33
10%-trimmed MSE	35.67	24.09	1.17
Features	86	selection of max. 10	41

3   Asset Prediction Model

Building upon the regression model of the previous section, in this section we present a stochastic model for answering the following question: *How many assets of type A are needed in the next n years?* The input parameters of this problem are:

- An asset type A. Associated with an asset type is a failure rate, such as MTBF (mean time between failure). MTBF is the expected time between failures of the asset. We only consider failures that cause the replacement of the asset.[5] The failures can be seen as random samples of a non-repairable population and the failure times follow a distribution with some probability density function (PDF).

- A scope $S = \{s_1, \ldots, s_n\}$ is a set of asset groups. Basically, each asset group corresponds to all assets of type A of an installed or planned system that must be taken into account for the forecast. Each member of an asset group must have the same installation date. As we will see later, this is important for computing the renewal numbers (older assets are more likely to fail than newer ones). Therefore, an installed system may be represented by more than one asset group: one group for all assets initially installed and still alive, and the other groups for additional assets which were installed afterwards due to necessary asset replacements.

Each $s \in S$ has the following properties:

- M_s is a random variable representing the number of assets in the asset group $s \in S$; $\Pr(M_s = n), n \in \mathbb{N}$, is the probability that s contains n elements.
- $tbos_s$: Begin of service time of all assets in $s \in S$. At this point in time the assets has been or will be installed in the field.[6]
- $teos_s$: End of service time of all assets in $s \in S$. This is the time when the service contract ends. After this point in time the assets need no longer be replaced when they fail.
- A probability $q_s \in [0, 1]$ representing the likelihood that the system containing $s \in S$ will be ordered. Trivially, for already installed assets $q_s = 1$.

- A point in time $t_{target} \in \mathbb{N}$ until that the prediction should be made. All assets in the scope whose service times overlap the period $[t_{now}, t_{target}]$ are to be taken into account. The service time of interest for each asset group $s \in S$ is then

$$\tau_s = min(teos_s, t_{target}) - tbos_s.$$

Algorithm 2. Total Asset Prediction. (Source: [33])

1. Input: (i) asset type A (ii) scope $S = \{s_1, \ldots, s_n\}$ (iii) target time t_{target}
2. Collect asset groups of given asset type and given scope
   (a) Existing assets from installed systems
   (b) Asset predictions for planned systems
3. Perform renewal asset analysis: $\hat{R}_s$
4. Compute estimator for each asset group: $\hat{N}_s$
5. Sum up asset group estimators: $\hat{N}_{total} = \sum_{s \in S} \hat{N}_s$

---

[5] A detailed differentiation of MTBF and MTTF (Mean Time To Failure) is beyond the scope of this paper.

[6] We use a simple representation of time: years started from some absolute zero time point. This simplifies arithmetic operations on time variables.

Please note that the computation of the forecast model is always done for assets of a given type A, so for the sake of simplifying the notation we omit to subscript all variables with A in this section.

The prediction of needed assets must take into account both renewal of already installed assets in case of non-repairable failures and new assets needed for planned systems. We will present a stochastic model that combines both cases. Algorithm 2 summarizes the steps how to compute such a prediction model. In step 1, the user has to provide an asset type, a scope, and a target point in time. In step 2, the asset groups from the installed base (existing projects) and the sales database (future projects) are collected. Then a stochastic model that represents needed renewal assets is computed based on failure rates/failure distributions of the assets and the service time periods of the asset groups. In step 4, an estimator is computed for each asset group using project order probability, the probability of the number of assets in the group, and the renewal model. Finally, all these estimators are summed up to an overall asset estimator $\hat{N}_{total}$. We present the details in the next subsections.

3.1   Renewal Processes

To estimate the number of assets needed for replacement of failing assets we resort to the theory of renewal processes – see, e.g., [6,18]. In this subsection, we recap the main definitions of renewal processes and show how we apply it to predict the number of assets that must be replaced because of failures. A *renewal process* is a stochastic model for renewal events that occur randomly in time. Let X_1 be the random variable representing the time between 0 and the first necessary renewal of an asset. The random variables $X_2, X_3, \ldots$ are the subsequent renewals. These variables X_i are called inter-arrival times. In our case, their distribution is directly connected to the failure rates or MTBF of the asset type. So $(X_1, X_2, X_3, \ldots)$ is a sequence of independent, identically distributed random variables representing the time periods between renewals. X_i takes values from $[0, \infty)$, and $\Pr(X_i > 0) > 0$. Let $f_X(t)$ be the PDF and $F_X(t) = \Pr(X \leq t)$ be the distribution function of the variables X_i.

The random variable T_n for some number $n \in \mathbb{N}$ represents the so-called arrival time; $F_{T_n}(t) = \Pr(T_n \leq t)$ represents the probability of n renewals up to time t. T_n is simply the sum of the inter-arrival variables X_i:

$$T_n = \sum_{i=1}^{n} X_i$$

The PDF of T_n is therefore the convolution of its constituents:

$$f_{T_n} = f_X^{*n} = f_X * f_X * \ldots * f_X$$

Remark. To add two random variables one has to apply the convolution operator on the their PDFs. The convolution operator $*$ for two functions f and g is defined as:

$$(f * g)(t) = \int f(t')g(t - t')\, dt' \text{ , continuous case}$$

$$(f * g)(n) = \sum_{m=-\infty}^{\infty} f(m)g(n - m) \text{ , discrete case}$$

The expression f^{*n} stands for applying the convolution operator on a function f n times. For many probability distribution families, summing up two random variables and therefore computing the convolution of their PDFs is easy. For instance, the sum of two Poisson distributed variables with parameters λ_1 and λ_2 is: $Poi[\lambda_1] + Poi[\lambda_2] = Poi[\lambda_1 + \lambda_2]$. (End of remark.)

In the renewal process, the arrival time variables T_n are used to create a random variable N_t that counts the number of expected renewals in the time period $[0, t]$. It is defined in the following way:

$$N_t = |\{n \in \mathbb{N} : T_n \leq t\}| \text{ for } t \geq 0$$

The arrival time process T_n and the counting process N_t are kind of inverse to each other, where the probability distribution of N_t can be derived from the probability distribution of T_n. Thus, the theory of renewal processes gives us a tool for deriving the counting variable N_t from a given failure distribution X of an asset. For complicated distributions of X, the derivation of N_t could get elaborate and could reasonably be done by numeric methods only, but it is straight-forward for some prominent distribution families. The most important case is the Poisson process, where X has an exponential distribution with parameter λ. In this case, the n-th arrival time variable T_n has a Gamma distribution with shape parameter n and rate parameter λ, and the counting variable N_t has a Poisson distribution with parameter λt.

3.2   Prediction Model

In this subsection we describe how to combine the number of assets in each asset group M_s $(s \in S)$, the above described renewal counting variable N_{τ_s}, and the probability q_s representing the likelihood that the assets will be ordered to an estimator, i.e., a probability mass function[7] (PMF), reflecting the number of needed assets in a given time period.

Definition 1 (Renewal Estimator $\hat{R}_s$). *Let M_s be a random variable representing the number of assets in the asset group $s \in S$. Let N_{τ_s} be the renewal counter for the service time period τ_s of asset group s. The* renewal estimator *$\hat{R}_s$ is a random variable, where $Pr(\hat{R}_s = n), n \in \mathbb{N}$, is the probability that exactly*

---

[7] A probability mass function (PMF) is the discrete counterpart of a probability distribution function (PDF). A PMF $f(n)$ corresponding to a random variable X is $Pr(X = n), n \in \mathbb{N}$.

n renewal assets are needed in total for the asset group $s \in S$. It's PMF $f_{\hat{R}_s}$ is defined in the following way (Source: [33]):

$$f_{\hat{R}_s}(n) := \sum_{k=0}^{\infty} f_{M_s}(k) f_{N_{\tau_s}}^{*k}(n) \tag{11}$$

Definition 2 (Asset Group Estimator $\hat{N}_s$). Let M_s be a random variable representing the number of assets in the asset group $s \in S$. Let q_s be the probability that the project containing the assets of s will be ordered. Let $\hat{R}_s$ be the renewal estimator as defined above. $\hat{N}_s$ is a random variable with $Pr(\hat{N}_s = n), n \in \mathbb{N}$, representing the probability that the number of assets needed for s is exactly n. It is the sum of the number of assets and the number of renewals scaled by the order probability. Its PMF is (Source: [33]):

$$f_{\hat{N}_s}(n) := q_s(f_{M_s}(n) * f_{\hat{R}_s}(n)) + (1 - q_s)\delta_0(n) \tag{12}$$

Remark. The delta function $\delta_k(x)$ (also called unit impulse) is 1 at $x = k$ and otherwise 0. We use the delta function to express certainty of zero assets in the case that the project is not ordered. The convolution with the delta function can be used for shifting: $f(x) * \delta_k(x) = f(x - k)$. (End of remark.)

Definition 3 (Total Asset Estimator $\hat{N}_{total}$). The total asset estimator $\hat{N}_{total}$ is a random variable with $Pr(\hat{N}_{total}n), n \in \mathbb{N}$, representing the probability that the total number of assets needed for all asset groups $s \in S$ until t_{target} is exactly n. It is the sum of the asset group estimators $\hat{N}_s$ (Source: [33]).

$$\hat{N}_{total} \sum_{s \in S}^{n} \hat{N}_s \tag{13}$$

Its PMF is the convolution of the PMFs of the asset group estimators $\hat{N}_s$ (Source: [33]):

$$f_{\hat{N}_{total}}(n) := \ast_{s \in S}\, f_{\hat{N}_s}(n) \tag{14}$$

The probability distribution $f_{\hat{N}_{total}}$ can now be used for calculating its expectation value, its variance or standard deviation, but also the number of needed assets with a guaranteed probability that the number will be high enough, i.e., compute the smallest n where $Pr(\hat{N}_{total} = n)$ is greater or equal some given probability, such as 0.75 or 0.95, corresponding to the certainty the forecast should provide.

The design of the variables in this framework is both suited for installed and planned systems. For a planned system, the probability that the system will be ordered is an information provided by sales experts. The estimation of how many assets will be needed can be predicted by the regression model described in Sect. 2. In this case, the estimator M_s corresponds to a random variable derived from the regression model $\hat{y} = f(\bar{x})$, where $\bar{x}$ is the feature vector of the planned system. See Eq. 9 in Sect. 2.

For asset groups of installed systems, the order probability q_s is simply 1, and the estimator variable M_s for the asset count is the delta function $\delta_k(n)$ with k being the actual number of installed assets. In this case the asset group estimator $\hat{N}_s$ is simply the shifted renewal estimator $\hat{R}_s$ with PDF $f_{\hat{R}_s}(n-k)$. It should be noted that $\hat{N}_{total}$ not only contains predicted assets, but also all already installed assets. These could be simply subtracted from $\hat{N}_{total}$ if only the number of assets to be ordered in the future are needed.

3.3 Example

Figure 5 shows a small example that shall demonstrate the combination of statistical methods of our asset prediction framework.

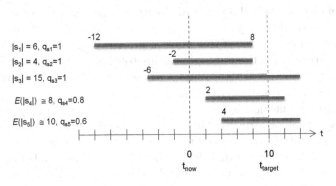

Fig. 5. Example of an asset prediction problem, containing 3 asset groups of existing projects (s_1, s_2, s_3) and 2 future projects with asset groups s_4 and s_5. (Source: [33]).

We use an exponential distribution of the failure rates of the assets with $\lambda = 0.125$ (i.e., 0.125 failures per year expected), corresponding to a MTBF of 8 years. So the renewal counting variables N_τ are Poisson distributed: $N_\tau \sim Poi(\lambda\tau)$. The number of assets of the existing systems 1 to 3 are 6, 4, and 15, so $M_{s_1} \sim \delta_6$, $M_{s_2} \sim \delta_4$, and $M_{s_3} \sim \delta_{15}$. We use a normal distribution for the number estimators M_{s_4} and M_{s_5} of the two planned systems with $M_{s_4} \sim \mathcal{N}(8, 1.74)$ and $M_{s_5} \sim \mathcal{N}(10, 1.74)$. The resulting asset group estimators and the total asset estimator are depicted in Fig. 6. Some interesting results are shown in Table 3. The first row shows the resulting expectation values of the number of assets. The other rows in Table 3 show the variances and standard deviations of our estimator probability functions, along with 3 examples of asset estimations with specified likelihood, i.e., $F^{-1}(p)$ is the quantile (i.e., the inverse cumulative probability) for a given probability value p and stands for the number of assets n with $\Pr(F \leq n) >= p$. While usually only this expectation value is used for asset prediction, our approach provides valuable additional information, like the standard deviation, and the possibility to find an asset number estimation with high reliability, like $F^{-1}(0.95)$.

Table 3. Some resulting probability values for the example scenario depicted in Fig. 5.

	s_1	s_2	s_3	s_4	s_5	Total
$\mathbb{E}$	21	9	45	13	11	99
var	14.91	4.97	29.68	52.26	80.79	181.84
σ	3.86	2.23	5.45	7.23	8.99	13.48
$F^{-1}(0.50)$	21	9	45	15	14	99
$F^{-1}(0.75)$	24	10	49	18	18	108
$F^{-1}(0.95)$	28	13	54	22	22	120

3.4 Advanced Visualization of Prediction Results

We have defined the overall number of hardware assets of a set of systems over the years as a random process, taking new systems and asset renewals into account. The representation of this random process results in the random variable $\hat{N}_{total}$, representing the total asset estimator. For a given year in the future (t_{target}), many useful prediction values, such as expectation value, variance, and confidence intervals, can be derived from $\hat{N}_{total}$.

An important question is how to present these results to the user. A first attempt was a tabular form similar to Table 3 together with a diagram similar to Fig. 6. A usability study revealed that customers had trouble to quickly grasp the meaning of the diagrams that depict the PMF with the number of predicted assets on the x-axis and the probability on the y-axis. Technical engineers are used to have a time-line on the x-axis.

So we switched to another representation of the asset predictor $\hat{N}_{total}$, as shown in Fig. 7. The x-axis is now a time-line, containing all years from the current year to a target year in the future specified by the customer. The y-axis represents the number of predicted assets. For each year, we inserted a 90° rotated PMF into the graph, containing also the expectation value and the quantile $F^{-1}(p)$ for a user-specified probability value p. The customer understands now quickly and easily all important information she/he expects: the expectation value, the quantile, and additionally the detailed probability distribution of the $\hat{N}_{total}$ for each year in the future.

Figure 7 shows the resulting total asset estimators of a real world example.[8] We applied our method to the data set used in Sect. 2.3 and computed predictors for the number of assets of type $A41$ for the years 2019 till 2027. About 50 installed systems with 670 instances of asset $A41$ and 10 future systems have been taken into account. Subtracting the already installed 670 assets, we get an expectation value of 541 additional assets for the year 2027, covering the assets for the 10 new systems and asset renewals. A prediction with 90% likelihood results in the need of 600 new assets.

---

[8] Although the example is based on real-world data, we had to change some numbers, MTBF, and order probabilities in order to not disclose business information. The data are realistic but do not reflect actual business data.

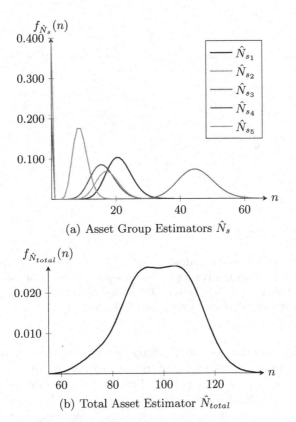

(a) Asset Group Estimators $\hat{N}_s$

(b) Total Asset Estimator $\hat{N}_{total}$

Fig. 6. Probability mass functions of predicted assets for asset groups and total asset estimator for the example scenario depicted in Fig. 5. (Source: [33])

4   Related Work

In industry, asset management is defined as the management of physical, as opposed to financial, assets [1]. Managing assets may comprise a broad range of different potentially overlapping objectives such as planning, manufacturing, and service. Depending on the scope of asset management, the data included determine the possibilities of predictive data analytics [12]. An essential part of asset management, especially in manufacturing, is to estimate the condition (health) of an assset.

In the last decade, an engineering discipline, called *prognostics*, has evolved aiming to predict the remaining useful life (RUL), i.e., the time at which a system or a component (asset) will no longer perform its intended function [26]. Since the reason for non-performance is most often a failure, data-driven prognostics approaches consist of modeling the health of assets and learning the RUL from available data [16,17]. [2] presents a stochastic method for data-driven RUL prediction of a complex engineering system. Based on the health value of assets,

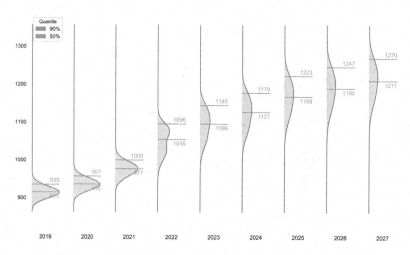

Fig. 7. An advanced graphical visualization of $\hat{N}_{total}$ for our example estimator of asset A41. The x-axis represents the years for which we predict the asset values. The y-axis represents the number of predicted assets. For each year, a 90° rotated PMF for $\hat{N}_{total}$ is shown, annotated by expectation value and 90% quantile.

logistic regression and the assessment output is used in a Monte Carlo simulation to estimate the remaining useful life of the desired system. In [11], the weighted RUL of assets is calculated by applying Principal Component Analysis (PCA) in combination with Restricted Boltzmann Machine (RBM) with the goal to optimize the productivity in a cyber-physical manufacturing system.

Other predictive methods aim for predicting the *obsolescence* of assets based on sales data. [21] proposes a data-mining approach including linear regression to estimate when an asset becomes obsolete. Based on this approach, time series models are used in [13] instead of linear regression models with the conclusion that prediction over years may be inaccurate. [8] examines the detection of obsolescence in a railway signaling system with a Markovian model and mentions that forecasting obsolescence of assets may be inexact. [25] drafts the potential application of predictive asset management in the railway domain by incorporating heterogeneous data sources such as construction and sales data. One of the main challenges is predictive asset management for long-term maintenance which coincides with our context.

One possible starting point to address the prediction of *long-term* asset management is formulating the product life cycle with evolutionary parametric drivers that describe an asset type whose performance or characteristics evolve over time [24]. Since evolutionary parametric drivers cannot be found in our database, procurement life modeling may be used [20]. The most prominent model of the mean procurement life, which is analogous to the mean-time-to-failure (MTTF), is the bathtub curve [10], usually modeled by Weibull distributions. This model represents three regimes in the lifetime of a hardware module: the first phase shows high but decreasing failure rates ("infant mortality"), the main, middle phase shows low, constant failure rates, and the third phase represents "wear out failures" and therefore increasing failure rates.

A prominent model in *preventive maintenance* (see, e.g., [5]) represents the replacement of an asset after a constant time before the failure probability gets too high. This can be modeled by a distribution function F that consists of two parts: the first part is some "conventional" distribution function, but from a defined replacement time t_r on, the probability of replacement is just 1.

Our input data is taken from the installed base (database of currently installed systems) and the sales database containing forecasts of expected future projects (planned systems). Unlike previous predictive obsolescence approaches where the main goal is to estimate the date when particular assets become obsolete [19], the combination of our data sets enables us to represent not only the numbers of renewal assets needed for n years but also the uncertainty of the estimation for long-term maintenance.

5   Conclusion and Future Work

In this paper, we proposed a predictive asset management method for hardware assets. The proposed method consists of two combined phases.

In the first phase, a regression model is learned from installed systems to predict assets for planned systems. Analyzing the performance of feature selection, it was worth following an extension of this phase since the method applied is potentially prone to distractions due to outliers. We integrated a robust method and observed results according to prediction accuracy. Experimental results of the feature selection process show a significant increase in performance.

In the second phase, a stochastic model is used for summing up all assets needed in the next n years for existing and also future projects, taking renewals of failing components into account. Results are presented in the course of sketched distribution functions which may be complex for interpretation. In order to adequately present results, an advanced visualization of prediction results supports technical engineers to quickly grasp all important information.

An important problem for future work is to find a minimal subset of features that is significant for predicting all assets. It would be interesting to take costs of feature measurement into account. E.g., in the railway domain it is easier to count signals and track switches than insulated rail joints of a railroad region. An optimal subset of features is one with minimal total measurement costs.

Presently, we use a given MTBF per asset type in our method. We assume that prediction accuracy can be further improved by applying advanced prognostics techniques for customizing the MTBF to the respective system context.

Although we tested our method only on data from rail automation, we suppose that applying it to hardware components in other domains will produce similar results of prediction. This needs to be evaluated in future work.

References

1. Amadi-Echendu, J., et al.: What is engineering asset management? In: Amadi-Echendu, J., Brown, K., Willett, R., Mathew, J. (eds.) Definitions, Concepts and Scope of Engineering Asset Management. Engineering Asset Management Review, vol. 1, pp. 3–16. Springer, London (2010). https://doi.org/10.1007/978-1-84996-178-3_1

2. Bagheri, B., Siegel, D., Zhao, W., Lee, J.: A stochastic asset life prediction method for large fleet datasets in big data environment. In: ASME 2015 International Mechanical Engineering Congress and Exposition, pp. V014T06A010–V014T06A010. American Society of Mechanical Engineers (2015)

3. Dixon, S.J., et al.: Pattern recognition of gas chromatography mass spectrometry of human volatiles in sweat to distinguish the sex of subjects and determine potential discriminatory marker peaks. Chemom. Intell. Lab. Syst. **87**(2), 161–172 (2007)

4. Filzmoser, P., Liebmann, B., Varmuza, K.: Repeated double cross validation. J. Chemom. **23**(4), 160–171 (2009)

5. Gertsbakh, I.: Reliability Theory: With Applications to Preventive Maintenance. Springer, New York (2013)

6. Grimmett, G., Stirzaker, D.: Probability and Random Processes. Oxford University Press, Oxford (2001)

7. Hoffmann, I., Serneels, S., Filzmoser, P., Croux, C.: Sparse partial robust M regression. Chemometr. Intell. Lab. Syst. **149**, 50–59 (2015)

8. Jenab, K., Noori, K., Weinsier, P.D.: Obsolescence management in rail signalling systems: concept and markovian modelling. Int. J. Prod. Qual. Manag. **14**(1), 21–35 (2014)

9. Jennings, C., Wu, D., Terpenny, J.: Forecasting obsolescence risk and product life cycle with machine learning. IEEE Trans. Compon. Packag. Manuf. Technol. **6**(9), 1428–1439 (2016)

10. Klutke, G., Kiessler, P.C., Wortman, M.A.: A critical look at the bathtubcurve. IEEE Trans. Reliab. **52**(1), 125–129 (2003). https://doi.org/10.1109/TR.2002.804492

11. Lee, J., Jin, C., Bagheri, B.: Cyber physical systems for predictive production systems. Prod. Eng. **11**(2), 155–165 (2017)

12. Li, J., Tao, F., Cheng, Y., Zhao, L.: Big data in product lifecycle management. Int. J. Adv. Manuf. Technol. **81**(1–4), 667–684 (2015)

13. Ma, J., Kim, N.: Electronic part obsolescence forecasting based on time series modeling. Int. J. Precis. Eng. Manuf. **18**(5), 771–777 (2017)

14. Maronna, R., Martin, R.D., Yohai, V.: Robust Statistics, vol. 1. Wiley, Chichester (2006). ISBN

15. Mooi, E., Sarstedt, M., Mooi-Reci, I.: Market Research. STBE. Springer, Singapore (2018). https://doi.org/10.1007/978-981-10-5218-7

16. Mosallam, A., Medjaher, K., Zerhouni, N.: Component based data-driven prognostics for complex systems: methodology and applications. In: 2015 First International Conference on Reliability Systems Engineering (ICRSE), pp. 1–7. IEEE (2015)

17. Mosallam, A., Medjaher, K., Zerhouni, N.: Data-driven prognostic method based on bayesian approaches for direct remaining useful life prediction. J. Intell. Manuf. **27**(5), 1037–1048 (2016)

18. Randomservices.org: Renewal processes (2017). http://www.randomservices.org/random/renewal/index.html. Accessed 3 Feb 2018

19. Sandborn, P.: Forecasting technology and part obsolescence. proceedings of the institution of mechanical engineers, part B: J. Eng. Manuf. **231**(13), 2251–2260 (2017)
20. Sandborn, P., Prabhakar, V., Ahmad, O.: Forecasting electronic part procurement lifetimes to enable the management of DMSMS obsolescence. Microelectron. Reliab. **51**(2), 392–399 (2011)
21. Sandborn, P.A., Mauro, F., Knox, R.: A data mining based approach to electronic part obsolescence forecasting. IEEE Trans. Compon. Packag. Technol. **30**(3), 397–401 (2007)
22. Serneels, S., Croux, C., Filzmoser, P., Van Espen, P.J.: Partial robust m-regression. Chemometr. Intell. Lab. Syst. **79**(1–2), 55–64 (2005)
23. Smit, S., Hoefsloot, H.C., Smilde, A.K.: Statistical data processing in clinical proteomics. J. Chromatogr. B **866**(1–2), 77–88 (2008)
24. Solomon, R., Sandborn, P.A., Pecht, M.G.: Electronic part life cycle concepts and obsolescence forecasting. IEEE Trans. Compon. Packag. Technol. **23**(4), 707–717 (2000)
25. Thaduri, A., Galar, D., Kumar, U.: Railway assets: a potential domain for big data analytics. Procedia Comput. Sci. **53**, 457–467 (2015)
26. Vachtsevanos, G.J., Lewis, F., Hess, A., Wu, B.: Intelligent Fault Diagnosis and Prognosis for Engineering Systems. Wiley, Hoboken (2006)
27. Wilcox, R.R., Keselman, H.: Modern robust data analysis methods: measures of central tendency. Psychol. Meth. **8**(3), 254 (2003)
28. Wold, H.: Nonlinear Estimation by iterative least squares procedures. In: David, F.N. (Hrsg.) Festschrift for J. Neyman: Research Papers in Statistics, London (1966)
29. Wold, H.: Model construction and evaluation when theoretical knowledge is scarce: theory and application of partial least squares. In: Kmenta, J., Ramsey, J.B. (eds.) Evaluation of Econometric Models, pp. 47–74. Elsevier, Amsterdam (1980)
30. Wold, S., Sjöström, M., Eriksson, L.: PLS-regression: a basic tool of chemometrics. Chemometr. Intell. Lab. Syst. **58**(2), 109–130 (2001)
31. Wurl, A., Falkner, A., Haselböck, A., Mazak, A.: Advanced data integration with signifiers: case studies for rail automation. In: Filipe, J., Bernardino, J., Quix, C. (eds.) DATA 2017. CCIS, vol. 814, pp. 87–110. Springer, Cham (2018). https://doi.org/10.1007/978-3-319-94809-6_5
32. Wurl, A., Falkner, A., Haselböck, A., Mazak, A.: Using signifiers for data integration in rail automation. In: Proceedings of the 6th International Conference on Data Science, Technology and Applications, vol. 1, pp. 172–179 (2017)
33. Wurl, A., Falkner, A.A., Haselböck, A., Mazak, A., Sperl, S.: Combining prediction methods for hardware asset management. In: Proceedings of the 7th International Conference on Data Science, Technology and Applications, DATA 2018, Porto, Portugal, pp. 13–23, 26–28 July 2018. https://doi.org/10.5220/0006859100130023

Linear vs. Symbolic Regression for Adaptive Parameter Setting in Manufacturing Processes

Sonja Strasser[1(✉)], Jan Zenisek[2], Shailesh Tripathi[1],
Lukas Schimpelsberger[1], and Herbert Jodlbauer[1]

[1] University of Applied Sciences Upper Austria, 4400 Steyr, Austria
{sonja.strasser, shailesh.tripathi,
lukas.schimpelsberger, herbert.jodlbauer}@fh-steyr.at
[2] University of Applied Sciences Upper Austria, 4232 Hagenberg, Austria
jan.zenisek@fh-hagenberg.at

Abstract. Product and process quality is playing an increasingly important role in the competitive success of manufacturing companies. To ensure a high quality level of the produced parts, the appropriate selection of parameters in manufacturing processes plays in important role. Traditional approaches for parameter setting rely on rule-based schemes, expertise and domain knowledge of highly skilled workers or trial and error. Automated and real-time adjustment of critical process parameters, based on the individual properties of a part and its previous production conditions, have the potential to reduce scrap and increase the quality. Different machine learning methods can be applied for generating parameter estimation models based on experimental data. In this paper, we present a comparison of linear and symbolic regression methods for an adaptive parameter setting approach. Based on comprehensive real-world data, collected in a long-term study, multiple models are generated, evaluated and compared with regard to their applicability in the studied approach for parameter setting in manufacturing processes.

Keywords: Process parameter setting · Manufacturing · Linear regression · Symbolic regression · Genetic programming

1 Introduction

Increasing customer requirements toward higher product and process quality force manufacturing companies to further improve their production processes in order to stay competitive [21, 28]. Especially in the area of high-tech manufacturing, even slight changes in the production environment, parameter setting of production facilities or raw material properties can lead to scrap or high effort in rework.

Quality is defined as the degree to which a commodity meets the requirements of the customer [6], whereby customers can be consumers of final products as well as other companies in a supply chain network. In case of suppliers in a supply chain, components, which are used as assembly parts in a final product, are delivered to a OEM (Original Equipment Manufacturer). For such parts, high quality requirements are defined, particularly concerning their specific dimensions, which have to be within

© Springer Nature Switzerland AG 2019
C. Quix and J. Bernardino (Eds.): DATA 2018, CCIS 862, pp. 50–68, 2019.
https://doi.org/10.1007/978-3-030-26636-3_3

predefined tolerances. There exist International Tolerance Grades of industrial processes, which identify what tolerances a given process can produce for a given dimension [12]. More precise production processes can decrease scrap rates or can even result in higher tolerance classes. As a consequence, a higher profit can be generated with these high quality parts.

The appropriate and prompt selection of process parameters in manufacturing processes plays a significant role to ensure the quality of the product, to reduce the machining cost and to increase the productivity of the process [20]. But the adjustment of process parameters to meet the quality requirements, e.g. dimensions of the part in predefined tolerances, can be a time-consuming and difficult task. Current approaches rely on rule-based schemes, expertise and domain knowledge of highly skilled workers or on trial and error. In the daily production routine of hard controllable production processes, a considerable amount of precious working time is spent for adjusting parameters. If a parameter setting results in a moderate average quality of the parts, it remains unchanged for the whole production lot, ignoring that each part has slightly varying properties and its own history in the previous processing steps.

As already slight variations of the product state during production can lead to costly and time-consuming rework or even scrap, [28] suggest an approach based on recording of the individual product's state along the entire production process. Whereas condition monitoring is mostly focused on a single manufacturing process, monitoring of the whole manufacturing programme has to be further investigated [3].

In [24] machine learning (ML) methods are applied to generate models based on experimental data, which automatically predict optimal parameters depending on the state of the produced part and its manufacturing conditions. This paper extends the approach for adaptive parameter setting in manufacturing processes introduced in [24] in two directions:

– In [24] linear regression and neural networks are selected as ML methods and applied in the case study. In this paper, we introduce symbolic regression for this approach because of its higher flexibility in comparison to linear regression and its higher interpretability in comparison to neural networks.
– The case study in [24] is based on data of 200 produced parts, collected in a single experiment. Now, we use data of more than 500 produced parts from four different experiments, which represent a higher variety of the production process.

Section 2 presents the related work for parameter setting in manufacturing processes. In Sect. 3 the methodology, consisting of the basic concept, data pre-processing and the application of linear regression and symbolic regression is explained. In the subsequent Sect. 4 we present the experimental data from a real-word manufacturing process and the results for all studied methods. The last Sect. 5 provides the concluding remarks and outlook.

2 Related Work

Existing work in the field of parameter optimization and parameter setting applies different methods and considers various manufacturing processes.

The most widespread machine learning method in this application field are artificial neural networks (ANN). In a review on artificial intelligence approaches for optimizing CNC machining parameters based on sensor information ANN is the dominating method. Also in the field of injection moulding various authors choose ANN for modeling the relationship between relevant input parameters (e.g. melt and mold temperature, pressure, cooling time) and quality measures (e.g. warpage, shrinkage) [10, 19, 23, 29]. For selecting optimal parameters, the trained networks are combined with heuristic optimization methods like genetic algorithms or particle swarm optimization.

In Additive Manufacturing ANN are applied to select optimal process parameters for different technologies in 3D printing (Fused Deposition Modeling, Selective Laser Melting or Sintering) [4, 7]. In a particleboard manufacturing process ANN model the relationship between process operating parameters and a critical strength parameter [5]. In a second step a genetic algorithm (GA) determines the process parameter values which result in the desired levels of the strength parameter.

In [25] a teaching-learning-based optimization algorithm is developed for process parameter optimization in a multi-pass turning operation. The authors state that this algorithm can be easily modified for parameter optimization of other manufacturing processes, such as casting, forming and welding. A combination of cluster analysis and support vector machines (SVM) is applied on product stat data along the manufacturing process to achieve the goal of improve quality monitoring [28]. A theoretical example is presented to illustrate the potential of the approach, but the application to a real-life manufacturing process is missing.

Linear regression and ANN are applied in a real-world case study, which illustrates an approach for adaptive parameter setting for manufacturing processes [24]. The results reveal that neural nets outperform linear regression on the training data, but application on the test data shows a significantly higher test error.

In an approach for estimating control parameters of a plasma nitriding process multiple different methods for generating a regression model are tested [15]. In addition to linear regression, SVM, Gaussian processes and random forest regression, genetic programming with offspring selection [1] and constants optimization [14] are used to evolve symbolic regression models. Based on the regression models, an inverse optimization problem is solved to get good combinations of parameters such that desired product qualities can be fulfilled simultaneously.

In this paper the approach introduced in [24] is applied on an extended data set collected in a long-term study in multiple experiments from a real-life industrial environment. As the target variable is a specific dimension of the produced part, different methods for regression (linear regression, symbolic regression and ANN) are chosen and compared.

3   Methodology

In this section, the approach for the setting of process parameters of a manufacturing process, introduced in [24], is summarized and extended by the application of symbolic regression. The goal is to determine adaptive process parameters for each individual produced part depending on its properties and previous manufacturing conditions. First, relevant process and product data and the basic concept are presented. Based on this concept, necessary pre-processing steps are formulated. Finally, the application of two machine learning models (linear regression and symbolic regression) for this parameter setting approach are discussed.

3.1 Basic Concept for Parameter Setting

In a multi-stage manufacturing process, a physical transformation of a product is performed in each step, so that, the raw material in step 1 is transformed into the final product in the last step N. In our approach, we consider only one specific product type, if there are multiple product types the developed concept has to be applied to each type separately.

The manufacturing data, which is necessary for the application of this concept consists of product and process data of the investigated production process. After each production step, all relevant product attributes have to be recorded, as well as all relevant process variables, describing the conditions under which the part was processed (see Fig. 1). Examples for product variables are the dimensions of a part (e.g. diameter, length and height), the weight or the density. Process variables describe the condition of the machine and the tools with which the part is processed (e.g. temperature, pressure, force or adjustments of the machine). At the beginning, it is not obvious, which variables will play an important role in this concept of parameter setting, so the policy is to record all available data and let the applied algorithms decide which variables to select.

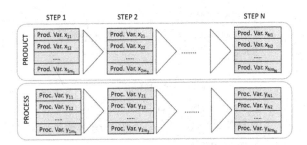

Fig. 1. Product and process variables [24].

The summarization of all process and product variables of one part describe its life cycle in the production process, which will be slightly different for each part. The reasons for this are variations in material properties and manufacturing conditions, also including environmental changes and influences of different human workers. A similar approach is introduced in [27] as product state based view.

Note that some of the variables can be chosen independently from others (e.g. type of raw material or adjustment of a certain process parameter) and some variables (e.g. quality relevant properties of the part) are dependent on the values of a set of other variables, although the precise relationship is usually unknown in practice.

In the approach, introduced in [24], the final objective is the determination of a suitable parameter setting of the last production step, adaptive for each part based on its manufacturing life cycle to fulfil the quality requirements. The concept is restricted to one process parameter u of the last step and a single quality measure z, which is one of the dependent product variables of step N. Figure 2 shows the two-step approach, where the first step models the relationship between product and process variables, including the selected process parameter, and the quality measure. Formally, this can be represented by a function f, returning an estimated value $\hat{z}$ for the quality measure:

$$\hat{z} = f(x_{ij}, y_{ij}, u) \tag{1}$$

In a second step the inverse function f^{-1} and the known optimal value for the quality measure are applied to provide an estimation for the process parameter $\hat{u}$, which is necessary to achieve the quality objective.

$$\hat{u} = f^{-1}(x_{ij}, y_{ij}, z) \tag{2}$$

Therefore, depending on the individual product and process variables of a part and the optimal value for the quality measure, which is the same for all parts of a certain product type, an adaptive parameter value can be determined. Different methods for modeling the function f in the first step of this approach are discussed in Sects. 3.3 and 3.4, after defining pre-processing tasks in Sect. 3.2.

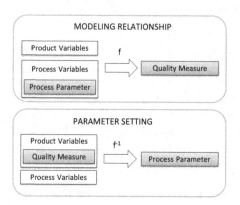

Fig. 2. Basic concept [24].

3.2   Data Pre-processing

Regardless of the applied methods for modeling the function f, necessary pre-processing steps are handling missing values and removing outliers. Depending on the machine learning methods, which are applied in the first step, additional pre-processing steps can be required.

In this article, we compare linear regression and symbolic regression. For both methods, collinearity of attributes causes problems. Perfect collinearity makes it impossible to invert matrices in some machine learning algorithms. High collinearity complicates the interpretation of selected variables and their coefficients, because there are multiple ways to model the same output. For detecting multicollinearity in the data, variance inflation factors (*VIF*) can be calculated. No collinearity is indicated by $VIF = 1$, whereas attributes with $VIF > 10$ should be discarded [13].

When we apply linear regression, we are restricted to numeric attributes. Nominal variables in the data set are replaced by binary attributes by dummy coding. A nominal variable with m levels has to be transformed in $m - 1$ dummy variables with 0 and 1 as possible values. On the other hand, for the employed symbolic regression implementation [26] no further pre-processing steps are necessary.

3.3 Application of Linear Regression

Building a linear regression model on all available process and product variables to predict the quality measure is likely to overfit the model to the data. So first, the variables, which have the biggest influence on the quality measures, have to be identified before they are used for building a linear regression model. From the point of view of the practitioners in the companies, it is desirable to get a model with a good accuracy, which only depends on a few variables. This would reduce cost and time for measuring and recording a huge amount of data from the production process. However, using too few variables will lead to bias.

Variable Selection. There are multiple methods for variable selection (see [9, 13, 17]), which can be applied, e.g. best subset selection, forward selection or backward elimination. Best subset selection fits a model to each combination of possible numbers of prediction variables. If there are p prediction variables, then 2^p models are trained and the best of them is selected. Because of the computational effort, the application is only possible, if p is not too high. Otherwise, stepwise methods, like forward selection and backward elimination, are alternatives, which only explore a restricted set of combinations. Forward selection starts with the best model containing only one variable and increases the number of variables in each step by one. Conversely, backward selection starts with all possible prediction variables and reduces the number in each step by one.

For all of these methods an evaluation criterion is needed for the comparison of different models in order to pick the best model [18]. Criteria based on the training data, like residual sum of squares (RSS) or the coefficient of determination (R^2) become better the more variables are included in the model. But, they are not suitable to assess the performance on new data. To gain insights on the performance of a model on unseen data, the test error can be estimated directly, e.g. by cross-validation or splitting the data randomly in training and test set. In this case suitable performance measures are the mean squared error (MSE) or the root mean squared error (RMSE):

$$MSE = \frac{1}{n_{test}} \sum_{i=1}^{n_{test}} (z_i - \hat{z}_i)^2 \qquad RMSE = \sqrt{MSE} \qquad (3)$$

Here the mean quadratic deviation of the estimated values $\hat{z}_i$ and the known target values z_i in the test set with n_{test} elements is calculated in the MSE. The RMSE makes it easier to interpret the test error because it is of the same unit as the target variable.

Another possibility is an indirect estimation of the test error by adjusting the training error to account for the bias due to overfitting. In [13] the Akaike information criterion (AIC), Bayesian information criterion (BIC), C_p value and Adjusted R^2 are suggested for this purpose. In the numerical results (see Sect. 4.2) Adjusted R^2 is used for the application:

$$Adjusted\ R^2 = 1 - \frac{RSS/(n-d-1)}{TSS/(n-1)} \qquad with$$

$$RSS = \sum_{i=1}^{n} (z_i - \hat{z}_i)^2, \quad TSS = \sum_{i=1}^{n} (z_i - \bar{z}_i)^2 \tag{4}$$

In this formula applied on a training set with n elements, the usual $R^2 = 1 - RSS/TSS$ is modified and takes the number of variables d into account. Including too much variables can lead to a decrease in Adjusted R^2. So the objective is to maximize this performance measure.

One of these criteria can be used for selecting an appropriate number of relevant variables for modeling. Here attention must be paid to ensure that the process parameter u, which has to be adjusted in the production process for each part, is included in the set of the selected variables.

Modeling and Parameter Setting. Once the variables are selected for building the regression model f, linear regression can be applied straight forward. In order to use all information available, the model can be trained on the whole data set. However, for estimating the performance, cross-validation or splitting training and test set has to be applied. By calculating the inverse function f^{-1} of the linear function and inserting the optimal value of the quality measure z and the individual product and process variables of a part, an estimation $\hat{u}$ for the parameter setting is yield.

3.4   Application of Symbolic Regression

The second employed modelling method is Genetic Programming, which generates so-called Symbolic Regression models within a stochastic, evolutionary process [16]. In contrast to most other machine learning models, which map input to output variables as non-transparent ensembles of trees or networks (black-box models), symbolic regression models are human interpretable mathematical functions in form of syntax trees (white-box models). Such white-box models pose potential for domain experts to gain a better understanding of the underlying modelled system, more than from sole analysis of correlations in black-box models. In comparison to the previously described linear regression models, genetic programming enables the identification of non-linear systems, which are represented by an extended mathematical vocabulary (e.g. arithmetic functions, logarithm, trigonometric functions etc.) in the resulting symbolic regression model, which makes them a superset of linear models. The approach has shown to be successful especially when used for modelling technical systems, in which mathematics constitute the natural description language [22]. Hence, it promises quite some potential for modelling industrial production processes.

In genetic programming, the model structure is developed automatically following a stochastic, evolutionary process [1]. Starting from a population of randomly created candidate solutions, the algorithm iteratively improves the models by randomly recombining (e.g. merging two functions) and mutating (e.g. removing or swapping symbols) the candidate solution's model-parts within a so-called generation. In reference to natural evolution's principle "survival of the fittest", the fitness of the models increases with the number of produced generations until the algorithm converges. The stochasticity within genetic programming makes it most likely that models from two different algorithm runs differ regarding the developed structure (i.e. genotypic representation) to some extent, although the models input response behaviour (i.e. phenotypic representation) can be very similar. This circumstance hampers the repeatability of experiments and the interpretability of the resulting models. Roughly speaking, there are numerous possible solutions with almost the same fitness quality [8], which is why this work provides more than just one final symbolic regression solution. For this paper we used the genetic programming based symbolic regression implementation in the open source framework HeuristicLab [26].

Variable Selection. To some extent feature selection is performed by the model creating process itself, since less useful variables are simply sorted out with progression of the evolutionary algorithm. However, supporting the algorithm concerning feature selection is highly recommended for several reasons: First, highly correlated variables pose another risk for generating models with different genotype but identical phenotype. Furthermore, the symbolic regression models created by the genetic programming algorithm are limited regarding their maximum size. This limitation is necessary to prevent the models from overfitting to the data and to prohibit overly complex functions. Due to the size restriction, tuples of highly correlated attributes with great impact on the target variable, which could be represented by a single variable, might cause that other variables cannot be considered. By performing the pre-processing steps, described in Sect. 3.2, we could effectively mitigate these risks.

Post-processing. In succession to the Genetic Programming algorithm, several semi-automatic routines have been applied on the resulting models in order to trim them regarding their interpretability and fitness quality. The syntax trees (i.e. mathematic formulas) have been simplified (by aggregating mathematical terms [2]), pruned (by replacing variables with only little impact on the model quality with constants), and optimized (by applying constant optimization [14]). All post-processing operations are implemented in HeuristicLab.

Modeling and Parameter Setting. In reference to the eventual aim of this work, the main advantage of symbolic regression compared with other machine learning model structures is that, as long as the resulting functions are mathematically invertible, they can be used for the task of parameter adaption as easily as described for linear regression models. All experiments have been repeated ten times by performing batch runs and the most promising models have been picked and post-processed. In Sect. 4, we present three genotypic different symbolic regression models with similar high fitness and reasonable complexity.

4 Experimental Results

In this section, the methods suggested in Sect. 3 are applied to manufacturing data collected from a real-world production process in metal processing industry, which consists of three production steps. The analyses concerning linear regression was implemented in R, whereas for symbolic regression and GP implementation Heuristic-Lab [26] was used to perform all experiments. This section starts with a detailed description of the available data. Then we present the results of each method separately. A comparison and discussion of the different methods concludes this section.

4.1 Experimental Data

In [24] experimental data of 200 produced parts and the associated process data was the input for the implemented case study. In this paper, we use data collected in a long-term study over a period of eight month recorded in four in distinct experiments. The case study data of [24] is included in this data as the first experiment. Overall, 13 product and 10 process variables of 527 parts are considered for the analysis (after removing some outliers due to obvious measurement errors).

According to Fig. 1, the following notation is used:

- x_{ij}: j-th product variables of step i
- y_{ij}: j-th process variable of step i

Table 1 shows the number of variables of each production step. The product variable of step 3 is an important quality measure, namely the height of the part ($z = x_{31}$) and the target variable of our regression analysis. The process variable of step 3 ($u = y_{33}$) is the process parameter, which should be selected adaptive for each part produced. Because most of the product and process variables are available after a regarded production step, less variables are available for step 3. Only variables, which are known in advance, like the adjustable process parameter, can be taken into account for a predictive model. y_{23} is the only nominal attribute in the data set, all the other variables are numeric. According to the suggested pre-processing steps in Sect. 3.2, it is replaced by four binary attributes (y_{231}, y_{232}, y_{233}, y_{234}), because 5 different values are possible.

There are three changes concerning the recorded variables in comparison to the case study in [24]:

1. In this paper, we restrict our analysis to quality measure 1, the height of the part. The diameter of the part, which was studied in [24] as the second quality measure is nearly constant for all parts. Its dimension is mainly determined by the tools used in the production process, whereas changes of the process parameter cause only slight variations.

2. The second process variable of step 3 (y_{32}), which was used in [24] is neglected in this study. It is a temperature, which is nearly constant under normal production conditions.

3. The process parameter y_{31}, which can be found in [24], has to be adjusted by a certain reference value in order to be comparable between different experiments. In order to define different variable names for different measures, we choose y_{33} for this adjusted version of the parameter, although it is the only process variable in step 3.

Table 1. Number of product and process variables.

Variables	Step 1	Step 2	Step 3	Total
Product variables	6	6	1	13
Process variables	4	4	1	9
Total	10	10	2	22

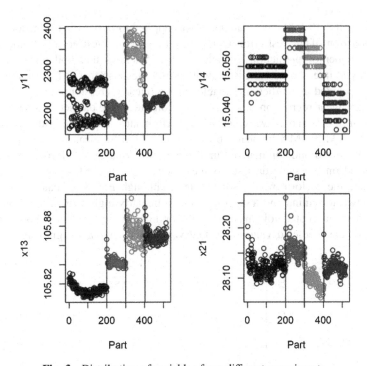

Fig. 3. Distribution of variables from different experiments.

The data collected in different experiments represents the high variety of the production process. Attribute values in a certain experiment are quite homogeneous, but there are major deviations between the experiences. Some distributions of the process and product variables are displayed in Fig. 3, where the vertical lines separate the four different experiment. This variety in the data raises the question whether linear regression is still an appropriate method for modeling the relationship between the input variables and the quality measure or if non-linear terms should be taken into account by applying symbolic regression.

4.2 Results for Linear Regression

Before modeling the relationship between product and process variables and the quality measure by a linear function, we check the input variables for multicollinearity. We calculate VIF, using the R package "car". The values range from $VIF(y_{33}) = 1.18$ to a maximum value of $VIF(x_{12}) = 8.06$, so no variables have to be eliminated due to their collinearity.

The moderate number of input variables allows us to use best subset selection for variable selection. For this purpose we use the "regsubsets" function from the R-package "leaps". Representative for the evaluation of the criteria mentioned in Sect. 3.3, Fig. 4 displays the results of adjusted R^2. Although the optimal number of prediction variables is 13, almost as good results of adjusted R^2 are already reached with five variables.

Additionally, 10-fold cross-validation was applied to best subset selection to get a direct estimation of the test error. Comparing training and test error for models with different number of prediction variables in Fig. 5 shows that both errors decrease significantly at five variables. This evaluation indicates also that at least five variables should be included as prediction variables in the linear regression model.

Now the linear regression model is trained on the whole dataset in order to obtain models with 5, 6, 8 and 10 variables. Their performance, measured with different key figures (R^2, RMSE, MAE: mean absolute error, MAPE: mean average percentage error), is evaluated and summarized in Table 2. Note that this performance measures are evaluated on the same data set as the models were trained. Therefore, they will overestimate the performance to some extent, although Fig. 5 indicates that the difference between training and test error is not particularly high. Comparing RMSE and MAE with the accepted tolerance of 0.06 for this quality measure shows that the deviation of the predicted values to actual values is on average significantly smaller

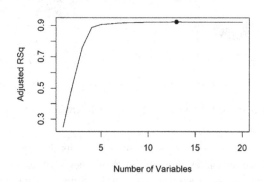

Fig. 4. Adjusted R squared for linear regression models.

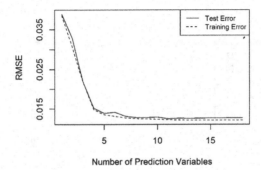

Fig. 5. Cross-validation of linear regression models.

than this tolerance. In all of the models the process parameter u, the product variables x_{12}, x_{21}, x_{26} and the process variable y_{14} are selected.

The formula obtained by linear regression for estimating quality measure 1 as a function of five prediction variables is:

$$\hat{z} = \hat{x}_{31} = f(u, x_{12}, y_{14}, x_{21}, x_{26})$$
$$= 57.476 - 0.245u + 0.434x_{12} - 4.641y_{14} + 0.484x_{21} + 0.020x_{26} \qquad (5)$$

Figure 6 provides a graphical comparison of actual and predicted values of the quality measure. For this figure, the parts are sorted ascending by their actual quality measure.

The optimal value of this quality measure is 27.29. Inserting this information in Eq. (5) and calculating the inverse function of f yields a function for estimating the process parameter u individually for each part, in dependence of three product variables and one process variable:

$$z = 27.29 \Rightarrow$$
$$\hat{u} = 123.208 + 1.77x_{12} - 18.943y_{14} + 1.976x_{21} + 0.082x_{26} \qquad (6)$$

Table 2. Validation of linear regression models.

Number of variables	5	6	8	10
Variables	u	u	u	u
	x_{12}, y_{14}	x_{12}, x_{13}, y_{14}	x_{12}, x_{13}, x_{16}, y_{14}	x_{12}, x_{13}, x_{16}, y_{11}, y_{14}
	x_{21}, x_{26}	x_{21}, x_{26}	x_{21}, x_{26}	x_{21}, x_{26}, y_{233}, y_{234}
R^2	0.907	0.913	0.920	0.923
RMSE	0.0135	0.0132	0.0125	0.0124
MAE	0.0109	0.0105	0.0101	0.0100
MAPE	0.0398%	0.0386%	0.0369%	0.0365%

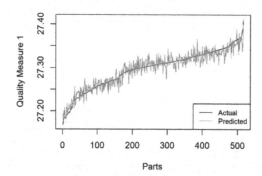

Fig. 6. Actual vs. predicted values using linear regression (5 variables).

4.3 Results for Symbolic Regression

For creating symbolic regression models besides basic arithmetic operations (+, −, *, /) and terminals (constants, variables), various others have been applied. Regarding the model size, we tested different restriction settings for the number of syntax tree nodes (i.e. mathematical symbols), to improve the interpretability of the generated models. For all experiments, we used genetic programming with offspring selection and constants optimization. With the permutations of following parameter options, experiments have been created and thus, a grid search performed.

- Symbolic Grammar: log and exp, trigonometric functions
- Max Tree Size: length = 25, 35, 50 nodes, depth = 100
- Population Size = 100, 250, 500

Each algorithm run was limited by a maximum of 1000 generations and a maximum selection pressure of 100. Furthermore, we configured a gender specific selector (female = proportional selection, male = random selection), a variety of alternating mutators (node type change, full and one-point shaker, remove and replace branch) with a probability of 15% and a simple subtree-crossover with a probability of 100%.

All experiments have been repeated ten times within batch runs. The best results have been achieved with simple symbolic operations (+, −, *, /), moderately sized tree length (length = 35 nodes) and employing a population size of 100 candidate solutions. The models with the highest fitness qualities have been picked and post-processed. The model parameters, the variables, selected by the GP-algorithm and the performance measured on the whole data set are provided in Table 3.

Table 3. Validation of symbolic regression models.

Model	M1	M2	M3
Model depth	8	5	5
Model length	24	19	18
Number of variables	8	10	8
Variables	u	u	u
	$x_{12}, x_{13}, x_{16}, y_{11}, y_{14}$	$x_{11}, x_{13}, x_{15},$	$x_{12}, x_{14}, x_{16},$
	x_{21}, x_{26}	y_{11}, y_{14}	y_{11}, y_{14}
		$x_{21}, x_{22}, x_{23},$	x_{21}, x_{22}
		x_{25}, x_{26}	
R^2	0.920	0.917	0.914
RMSE	0.0125	0.0128	0.0131
MAE	0.0101	0.0100	0.0105
MAPE	0.0369%	0.0368%	0.0386%

The first model M1 selects five product and two process variables in addition to the adjustable process parameter u. The structure of the developed function still enables the calculation of the inverse function for the determination of u, when we insert the optimal quality measure z:

$$\hat{z} = \hat{x}_{31} = f(u, x_{12}, x_{13}, x_{16}, y_{11}, y_{14}, x_{21}, x_{26})$$

$$= c_0 x_{16} + c_1 u + c_2 x_{12} + c_3 x_{21} +$$

$$(c_4 x_{13} + c_5 y_{14})\left(c_6 x_{13} + \frac{c_7 x_{12} + c_8 y_{11} + c_9 x_{26}}{c_{10} u + c_{11} x_{21} + c_{12} x_{16}}\right) c_{13} + c_{14}$$

$$
\begin{aligned}
c_0 &= 0.033 & c_8 &= 0.010 \\
c_1 &= -0.211 & c_9 &= -1.078 \\
c_2 &= 0.184 & c_{10} &= 1.235 \\
c_3 &= 0.403 & c_{11} &= -3.148 \\
c_4 &= 0.031 & c_{12} &= 0.769 \\
c_5 &= -1.606 & c_{13} &= 0.084 \\
c_6 &= 0.271 & c_{14} &= 42.034 \\
c_7 &= -3.498 &
\end{aligned}
$$

(7)

The second models M2 selects more variables than the first model, but its structure is nearly a linear function:

$$\hat{z} = \hat{x}_{31} = f(u, x_{11}, x_{13}, x_{15}, y_{11}, y_{14}, x_{21}, x_{22}, x_{23}, x_{25}, x_{26}) =$$

$$c_0 x_{23} + c_1 x_{22} + c_2 x_{21} + c_3 x_{26} + c_4 x_{25} + c_5 u + c_6 x_{11} + c_7 x_{13} + c_8 y_{14} x_{23} + \frac{c_9 y_{11}}{c_{10} x_{15}} + c_{11}$$

$$
\begin{aligned}
&c_0 = 0.529 &\qquad &c_6 = 0.087 \\
&c_1 = 0.144 & &c_7 = -0.520 \\
&c_2 = 0.484 & &c_8 = -0.034 \\
&c_3 = 0.020 & &c_9 = -0.001 \\
&c_4 = 0.109 & &c_{10} = 1.515 \\
&c_5 = -0.243 & &c_{11} = 50.749
\end{aligned}
\tag{8}
$$

The third model M3 is the simplest model presented in Table 3 with a slightly worse performance. Apart from one product of variables, all other terms are linear:

$$
\begin{aligned}
\hat{z} = \hat{x}_{31} &= f(u, x_{12}, x_{14}, x_{16}, y_{11}, y_{14}, x_{21}, x_{22}) \\
&= c_0 y_{14} + c_1 x_{12} + c_2 y_{11} + c_3 x_{22} + c_4 x_{16} + c_5 x_{21} + c_6 x_{14} u + c_7 \\
&c_0 = -4.263 \qquad\qquad c_4 = 0.0233 \\
&c_1 = 0.293 \qquad\qquad\quad c_5 = 0.357 \\
&c_2 = -8.3E - 05 \qquad\; c_6 = -0.017 \\
&c_3 = 0.096 \qquad\qquad\quad c_7 = 56.452
\end{aligned}
\tag{9}
$$

The estimated results from M3 are compared to the actual values of the quality measure in Fig. 7.

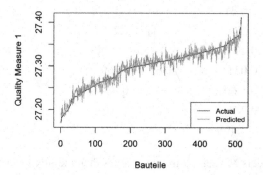

Fig. 7. Actual vs. predicted values using symbolic regression (M3).

4.4 Comparison of the Results

For comparison of the performance of linear regression and symbolic regression models, we validate multiple performance measures by a 10-fold cross-validation on the available data set. The average values of test errors are stated in Table 4. By analogy with [24], we also evaluated ANN for benchmarking. A neural net with one hidden layer was trained, whereby the optimal number of neurons in the hidden layer was selected through 10-fold cross-validation. As already stated in [24], neural nets are not suitable for this parameter setting approach due to the lack of model interpretability and impossibility of easily inverting the modeled function f. Although the coefficient of determination for ANN is better than for the other methods, the measures, describing the mean deviation of actual and predicted values, are higher for ANN. The performance of linear regression and symbolic regression is similarly, although linear regression performs always slightly better. Since the mathematical vocabulary used in linear regression models is a subset of the symbolic regression grammar, applying more extensive parameter tuning and larger scaled experiments, should eventually enable to find at least identic results when using genetic programming. Similarly, ANN are capable of approximating any arbitrary continuous function within a compact set when using at least one hidden layer, provided that parameters are tuned [11]. However, for the real-world application, the prediction quality differences are negligibly. The validation results indicates that the assumption of a linear relationship is still valid for this extended data set.

Table 4. Test errors of 10-fold cross-validation.

Performance	Lin Reg	Sym Reg	ANN
R^2	0.914	0.901	0.933
RMSE	0.0128	0.0140	0.0458
MAE	0.0103	0.0110	0.0365
MAPE	0.0375%	0.0402%	0.0881%

5 Conclusions

In manufacturing processes, the selection of appropriate process parameters contributes to an improvement of product quality and reduces scrap. Since every single part produced has slightly different properties and manufacturing conditions, an adaptive parameter setting, considering each parts history in the manufacturing life cycle, is required. In order to set up an automated and real-time process for parameter setting, based on machine learning methods, the relationship between relevant input variables and a certain quality measure has to be modeled.

In this paper we investigate different regression methods for the adaptive parameter setting approach suggested in [24]. Next to linear regression, we introduce symbolic regression models generated by genetic programming. Advantages of symbolic regression is the integrated variable selection and the good interpretability of the models.

For testing and comparing these different approaches we used data from a real-world manufacturing process, collected in a long-term study to cover a greater variety of the production conditions. Linear regression models with different number of variables were trained, whereby the most relevant variables were selected by a cross-validated best subset selection. Different parameters settings in the genetic programming algorithm provided symbolic regression models with varying complexity. To compare performance of different methods, average test errors from 10-fold cross-validation, based on different measures (RMSE, MAE, MAPE, R^2), are reported.

Summarizing the results, it can be stated, that for the available data set linear regression is a good choice for modeling the relationship of process and product variables and the quality measure. The generated linear function can easily be inverted to determine estimates for the process parameter, adaptive for each processed part. The performance of symbolic regression model is comparable to the linear models, but not better in our calculation runs. An interesting further research topic would be the comparison of these machine learning methods in other manufacturing processes. Another extension with practical relevance is the extension of this approach to multiple quality measures, which are weighted in an objective function.

Acknowledgements. The authors gratefully acknowledge financial support with the projects ADAPT and BAPDEC, which are funded by the country of Upper Austria in their program "Innovative Upper Austria 2020" and the project "Smart Factory Lab", which is funded by the European Fund for regional development (EFRE) and the country of Upper Austria as part of the program "Investing in Growth and Jobs 2014–2020".

Europäische Union Investitionen in Wachstum & Beschäftigung. Österreich.

References

1. Affenzeller, M., Winkler, S., Wagner, S., Beham, A.: Genetic Algorithms and Genetic Programming. Modern Concepts and Practical Applications. Numerical Insights, vol. 6. CRC Press, Boca Raton (2009)
2. Affenzeller, M., Winkler, S.M., Kronberger, G., Kommenda, M., Burlacu, B., Wagner, S.: Gaining deeper insights in symbolic regression. In: Riolo, R., Moore, Jason H., Kotanchek, M. (eds.) Genetic Programming Theory and Practice XI. GEC, pp. 175–190. Springer, New York (2014). https://doi.org/10.1007/978-1-4939-0375-7_10
3. Choudhary, A.K., Harding, J.A., Tiwari, M.K.: Data mining in manufacturing. A review based on the kind of knowledge. J. Intell. Manuf. **20**(5), 501–521 (2009). https://doi.org/10.1007/s10845-008-0145-x
4. Collins, P.C., et al.: Progress toward an integration of process–structure–property–performance models for "Three-Dimensional (3-D) Printing" of titanium alloys. JOM **66**(7), 1299–1309 (2014). https://doi.org/10.1007/s11837-014-1007-y
5. Cook, D.F., Ragsdale, C.T., Major, R.L.: Combining a neural network with a genetic algorithm for process parameter optimization. Eng. Appl. Artif. Intell. **13**(4), 391–396 (2000). https://doi.org/10.1016/S0952-1976(00)00021-X
6. DIN EN ISO 9001:2015: Quality management systems - Fundamentals and vocabulary (ISO 9000:2015) (2015)

7. Ding, D., et al.: Towards an automated robotic arc-welding-based additive manufacturing system from CAD to finished part. Comput. Aided Des. **73**, 66–75 (2016). https://doi.org/10.1016/j.cad.2015.12.003

8. Gustafson, S., Burke, E.K., Krasnogor, N.: On improving genetic programming for symbolic regression. In: The 2005 IEEE Congress on Evolutionary Computation. IEEE CEC 2005, Edinburgh, Scotland, UK, 02–05 September 2005, pp. 912–919. IEEE, Piscataway (2005). https://doi.org/10.1109/cec.2005.1554780

9. Guyon, I.: Feature Extraction. Foundations and Applications. Studies in Fuzziness and Soft Computing, vol. 207. Springer, New York (2006). https://doi.org/10.1007/978-3-540-35488-8

10. Hasan, K., Babur, O., Tuncay, E.: Warpage optimization of a bus ceiling lamp base using neural network model and genetic algorithm. J. Mater. Process. Technol. **169**(2), 314–319 (2005). https://doi.org/10.1016/j.jmatprotec.2005.03.013

11. Hornik, K.: Approximation capabilities of multilayer feedforward networks. Neural Networks **4**(2), 251–257 (1991). https://doi.org/10.1016/0893-6080(91)90009-T

12. ISO 286-1:2010: Geometrical product specifications (GPS)—ISO code system for tolerances on linear sizes (2010)

13. James, G., Witten, D., Hastie, T., Tibshirani, R.: An Introduction to Statistical Learning: With Applications in R. STS, vol. 103. Springer, New York (2013). https://doi.org/10.1007/978-1-4614-7138-7

14. Kommenda, M., Kronberger, G., Winkler, S., Affenzeller, M., Wagner, S.: Effects of constant optimization by nonlinear least squares minimization in symbolic regression. ACM (2013). http://dl.acm.org/ft_gateway.cfm?id=2482691&type=pdf

15. Kommenda, M., Burlacu, B., Holecek, R., Gebeshuber, A., Affenzeller, M.: Heat treatment process parameter estimation using heuristic optimization algorithms. In: Affenzeller, M., Bruzzone, A.G., Jimenez, E., Longo, F., Merkuryev, Y., Zhang, L. (eds.) Proceedings of the European Modeling and Simulation Symposium, pp. 222–227 (2015)

16. Koza, J.R.: Genetic Programming: On the Programming of Computers by Means of Natural Selection. MIT Press, Cambridge, MA, USA (1992). http://mitpress.mit.edu/books/genetic-programming

17. Miller, A.J.: Subset selection in regression. Monographs on Statistics and Applied Probability, vol. 95, 2nd edn. Chapman & Hall/CRC, Boca Raton (2002)

18. Murtaugh, P.A.: Methods of variable selection in regression modeling. Commun. Stat. Simul. Comput. **27**(3), 711–734 (2010). https://doi.org/10.1080/03610919808813505

19. Ozcelik, B., Erzurumlu, T.: Comparison of the warpage optimization in the plastic injection molding using ANOVA, neural network model and genetic algorithm. J. Mater. Process. Technol. **171**(3), 437–445 (2006). https://doi.org/10.1016/j.jmatprotec.2005.04.120

20. Pawar, P.J., Rao, R.V.: Parameter optimization of machining processes using teaching–learning-based optimization algorithm. Int. J. Adv. Manuf. Technol. **67**(5), 995–1006 (2013). https://doi.org/10.1007/s00170-012-4524-2

21. Robinson, C.J., Malhotra, M.K.: Defining the concept of supply chain quality management and its relevance to academic and industrial practice. Int. J. Prod. Econ. **96**(3), 315–337 (2005). https://doi.org/10.1016/j.ijpe.2004.06.055

22. Schmidt, M., Lipson, H.: Distilling free-form natural laws from experimental data. Science (New York, N.Y.) **324**(5923), 81–85 (2009). https://doi.org/10.1126/science.1165893

23. Shen, C., Wang, L., Li, Q.: Optimization of injection molding process parameters using combination of artificial neural network and genetic algorithm method. J. Mater. Process. Technol. **183**(2), 412–418 (2007). https://doi.org/10.1016/j.jmatprotec.2006.10.036
24. Strasser, S., Tripathi, S., Kerschbaumer, R.: An approach for adaptive parameter setting in manufacturing processes. In: Proceedings of the 7th International Conference on Data Science, Technology and Applications, Porto, Portugal, pp. 24–32. SCITEPRESS - Science and Technology Publications (2018). https://doi.org/10.5220/0006894600240032
25. Venkata Rao, R., Kalyankar, V.D.: Multi-pass turning process parameter optimization using teaching–learning-based optimization algorithm. Scientia Iranica **20**(3), 967–974 (2013). https://doi.org/10.1016/j.scient.2013.01.002
26. Wagner, S., et al.: Architecture and design of the HeuristicLab optimization environment. In: Klempous, R., Nikodem, J., Jacak, W., Chaczko, Z. (eds.) Advanced Methods and Applications in Computational Intelligence. Topics in Intelligent Engineering and Informatics, vol. 6, pp. 197–261. Springer, Heidelberg (2014). https://doi.org/10.1007/978-3-319-01436-4_10
27. Wuest, T., Klein, D., Thoben, K.-D.: State of steel products in industrial production processes. Procedia Eng. **10**, 2220–2225 (2011). https://doi.org/10.1016/j.proeng.2011.04.367
28. Wuest, T., Irgens, C., Thoben, K.-D.: An approach to monitoring quality in manufacturing using supervised machine learning on product state data. J. Intell. Manuf. **25**(5), 1167–1180 (2014). https://doi.org/10.1007/s10845-013-0761-y
29. Xu, Y., Zhang, Q., Zhang, W., Zhang, P.: Optimization of injection molding process parameters to improve the mechanical performance of polymer product against impact. Int. J. Adv. Manuf. Technol. **76**(9), 2199–2208 (2015). https://doi.org/10.1007/s00170-014-6434-y

Graph Pattern Index for Neo4j Graph Databases

Jaroslav Pokorný[1(✉)], Michal Valenta[2], and Martin Troup[2]

[1] Faculty of Mathematics and Physics, Charles University,
Prague, Czech Republic
pokorny@ksi.mff.cuni.cz
[2] Faculty of Information Technology, Czech Technical University,
Prague, Czech Republic
{valenta, troupmar}@fit.cvut.cz

Abstract. Nowadays graphs have become very popular in domains like social media analytics, healthcare, natural sciences, BI, networking, etc. Graph databases (GDB) allow simple and rapid retrieval of complex graph structures that are difficult to model in traditional information systems based on a relational DBMS. GDB are designed to exploit relationships in data, which means they can uncover patterns difficult to detect using traditional methods. We introduce a new method for indexing graph patterns within a GDB modelled as a labelled property graph. The index is based on so called graph pattern trees of variations and stored in the same database where the database graph. The method is implemented for Neo4j GDB engine and analysed on three graph datasets. It enables to create, use and update indexes that are used to speed-up the process of matching graph patterns. The paper provides details of the implementation, experiments, and a comparison between queries with and without using indexes.

Keywords: Graph databases · Indexing patterns · Graph pattern ·
Graph database schema · Neo4j

1 Introduction

A *graph database* (GDB) is a database that uses the graph structure to store and retrieve data. A GDB embraces relationships as a core aspect of its data model. The model is built on the idea that even though there is value in discrete information about entities, there is even more value in the relationships between them. Relaxing usual DBMS features, a native GDB can be any storage solution where connected elements are linked together without using an index (a property called *index-free adjacency*).

Similarly to traditional databases, we will use the notion of a *graph database management system* (GDBMS). GDBMSs proved to be very effective and suitable for many data handling use cases. For example, specifying a graph pattern and a set of starting points, it is possible to reach an excellent performance for local reads by traversing the graph starting from one or several root nodes, and collecting and aggregating information from nodes and edges. On the other hand, GDBMSs have their limitations [5]. For example, they are usually not consistent, since have very restricted

C. Quix and J. Bernardino (Eds.): DATA 2018, CCIS 862, pp. 69–90, 2019.
https://doi.org/10.1007/978-3-030-26636-3_4

tools to ensure a consistency. This property is typical for NoSQL databases [13], where GDBMs are often included.

A GDB can contain one (large) graph G or a collection of small to medium size graphs. The former includes, e.g., graphs of social networks, Semantic Web, geographical databases, the latter is especially used in scientific domains such as bioinformatics and chemistry or datasets like DBLP. Thus, the goal of query processing is, e.g., to find all subgraphs of G that are the same or similar to the given query graph. We can consider shortest path queries, reachability queries, e.g., to find whether a concept subsumes another one in an ontological database, etc. The query processing over a graph collection involves, e.g., finding all graphs in the collection that are similar to or contain subgraphs similar to a query graph. We focus on the first category of GDBs in this paper.

Graph search occurs in application scenarios, like recommender systems, analyzing the hyperlinks in WWW, complex object identification, software plagiarism detection, or traffic route planning. Gartner[1] believes that over 70% of leading companies will be piloting a GDB by 2018.

One of the most fundamental problems in graph processing is pattern matching. Specifically, a *pattern match query* searches over a G to look for the existence of a pattern graph in G. This problem can be expressed in the different graph data models as Resource Description Framework, property graphs as well as in the relational model. A property subclass of property graphs can even be modelled using XML documents. We will focus on general property graphs in this paper. Both above mentioned GDB types, however, reduce exact query matching to the subgraph isomorphism problem, which is NP-complete [15], meaning that this querying is intractable for large graphs in the worst-case. In context of Big Data we talk about Big Graphs [6]. Their storage and processing require special technics.

An effective implementation of each DBMS highly depends on the existence and usage of indexes. Nowadays, some effective indexes for nodes and edges already exist in GDB implementations (see, e.g., the evaluation [3] mentioned in Sect. 2.1), while structure-based indexes, which may be very useful for subgraph queries and for relationship-based integrity constraints checking, are yet rather the subject of research as it is described in Sect. 2. Particularly, there already exist indexing methods for (various kinds of) graph pattern matching, see, e.g., works [1, 14, 16].

In the paper, we focus on Neo4j GDBMS[2] and its possibilities to express an index of graph patterns. Neo4j is an open-source native GDBMS for storing and managing of property graphs, that offers functionality similar to traditional RDBMSs such as a declarative query language Cypher[3], full transaction support, availability, and scalability through its distributed version [10]. Cypher commands use partially SQL syntax and are targeted at ad hoc queries over the graph data. They enable also to create graph

---

[1] https://www.gartner.com/doc/3100219/making-big-data-normal-graph, last accessed 2018/11/14.

[2] https://neo4j.com/, last accessed 2018/11/14.

[3] http://neo4j.com/developer/cypher-query-language/, last accessed 2018/11/14.

nodes and relationships. Our goal is to extend the Cypher with new functionality supporting more efficient processing graph pattern queries.

Our work is an extension of the paper [8] that introduces a new approach to graph pattern indexing in Neo4j graph database environment. In this paper we present details of implementation, new experiments, and their discussion. The rest of the paper is organized as follows. In Sect. 2 we summarize some related works divided into two categories of graph indexing methods: value-based indexing and structure-based indexing. Section 3 introduces a GDB model based on (labelled) property graphs. We continue with graph pattern indexing and the details of the new method based on so called graph pattern trees of variations. An implementation is described in Sect. 4 and related experiments in Sect. 5. Section 6 gives the conclusion.

2   Background and Related Works

In general, graph systems use various graph analytics algorithms supporting with finding graph patterns, e.g., connected components, single-source shortest paths, community detection, triangle counting, etc. Triangle counting is used heavily in social network analysis. It provides a measure of clustering in the graph data which is useful for finding communities and measuring the cohesiveness of local communities in social network websites like LinkedIn or Facebook. In Twitter, three accounts who follow each other are regarded as a triangle.

One theme in graph querying is graph data mining finding frequent patterns. Frequent graph patterns are subgraphs that are found from a collection of graphs or a single massive graph with a frequency no less than a user-specified support threshold. Subgraph matching operations are heavily used in social network data mining operations.

Indexing is used in GDBs in many different contexts. Due to the existence of properties values in a GDB, graph indexes are of two kinds, in principle: *structure-aware* and *property-aware*. They occur in GDBMS in various forms from a fulltext querying support over indexing nodes, edges, and property types/values to indexes based on indexing non-trivial subgraphs.

2.1   Value-Based Indexing

Authors of [3] compare indexing used in two favourite GDBMSs – Neo4j and OrientDB[4]. The Cypher language of Neo4j enables to create indexes on one or more properties for all nodes that have a given label. OrientDB supports five classes of indexing algorithms: SB-Tree, HashIndex, Auto Sharding Index, and indexing based on the Lucene Engine (for fulltext and spatial data). SB-tree [4] is based on B-Tree with several optimizations related to data insertion and range queries. In Auto Sharding Index (key, value) pairs are stored in a distributed hash table.

---

[4] http://orientdb.com/orientdb, last accessed 2018/11/14.

Another native GDBMS Sparksee[5] uses B+-trees and compressed bitmap indexes to store nodes and edges with their properties. Titan[6] supports two different kinds of indexing to speed up query processing: graph indexes and node-centric indexes. Graph indexes allow efficient retrieval of nodes or edges by their properties for sufficiently selective conditions. Node-centric indexes are local index structures built individually per node. In large graphs, nodes can have thousands of incident edges.

2.2 Structure-Based Indexing

The design principle behind a structural index is to extract and index structural properties of database graphs, typically at insertion time, and use them to filter the search space rapidly in response to a query. Previous works have mainly focused on mining "good" substructure features for indexing. A good feature set improves the filtering power by reducing the number of candidate graphs, which leads to a reduction in the number of subgraph isomorphism tests in the verification step. Subtree features are also mined for indexing, and they are less time-consuming to be mined in comparison with more general subgraph features. Many methods take a path as the basic indexing unit. For example, the SPath algorithm [19] is centred on a local path-based indexing technique for graph nodes and transforms a query graph into a set of the shortest paths in order to process a query. The work [12] distinguishes three types of structure-based indexes: path-based index, subgraph-based index, and spectral methods.

It is remarkable, that different graph index structures have been used for different kinds of substructure features, but no index structure is enabled to support all kinds of substructure features. Authors of [18] propose a Lindex, a graph index, which indexes subgraphs contained in database graphs. Nodes in Lindex represent key-value pairs where the key is a subgraph in a GDB and the value is a list of database graphs containing the key. Frequent subgraphs are used for indexing in gIndex [16]. An introduction to graph substructure search, approximate substructure search and their related graph indexing techniques, particularly feature-based graph indexing can be found in [17]. In [20], the authors introduce a structure-aware and attribute-aware index to process approximate graph matching in a property graph.

A detailed discussion of different types of graph queries and a different mechanism for indexing and querying GDBs can be found in [11].

3 Modelling of Graph Databases

Although GDBMS can be based on various graph types, we will use a (*labelled*) *property graph model* whose basic constructs include:

- entities (nodes),
- properties (attributes),
- labels (types),

---

[5] http://www.sparsity-technologies.com/, last accessed 2018/11/14.

[6] http://titan.thinkaurelius.com, last accessed 2018/11/14.

- relationships (edges) having a direction, start node, and end node,
- identifiers.

Entities and relationships can have any number of properties, nodes and edges can be tagged with labels. Both nodes and edges are defined by a unique identifier (Id). Properties are of form key:domain, i.e. only single-valued attributes are considered. In graph-theoretic notions we also talk about *labelled and directed attributed multigraphs* in this case. It means the edges of different types can exist between two nodes. These graphs are used both for a GDB and its database schema (if any). In practice, this definition is not strictly enforced. There are GDBMSs supporting more complex property values, e.g. the already mentioned Neo4j.

When retrieving data from a GDB, one may want to query not only single nodes or relationships, but also more complex units consisting of these basic elements. Such units, *graph patterns*, can contain valuable information for many use cases. The fact that the graph can easily express such information is one of the main benefits of using such data model. Thus graph pattern matching is one of the key functionalities GDBs usually provide. In Sect. 3.1 we discuss shortly graph patters definable in the Cypher language and two basic methods for their indexing. Section 3.2 focuses on so called graph pattern trees of variations appropriate for organizing variations of a single graph pattern. Updating the index after performing DML operations is described in Sect. 3.3.

3.1 Graph Patterns

A wide variety of graph patterns can be found across different GDBs. Graph patterns have different information value that is based on type of data stored within a database and use cases that involve these graph patterns.

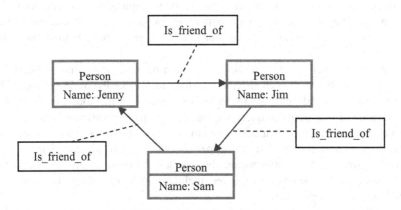

Fig. 1. Example of triangle [8].

One of widely used graph patterns, defined as GP = $(V_p; E_p)$, where $V = \{v_1; v_2; v_3\}$ and $E = \{(v_1; v_2); (v_2; v_3); (v_3; v)\}$, is called a *triangle*. In Cypher, a triangle can be expressed in a few different ways, but preferably, e.g., as

$$(n_1) - [r_1] - (n_2) - [r_2] - (n_3) - [r_3] - (n_1)$$

i.e., triangle patterns look for three nodes adjacent to each other regardless of edge orientation. That is, a direction can be ignored at query time in Cypher, i.e., the database graph behind can be handled as bidirectional. Figure 1 shows a triangle coming from a social graph. To retrieve such pattern using Cypher is easy for Neo4j. The problem arises when we focus only on structural features of the graph and want, e.g., all such triangles of people with their friendship. Then a structure-based index can be helpful.

A graph pattern index is basically a data structure that stores pointers that reference graph pattern units within the database. Indexes can be either stored in the same database as the actual data or in any external data store. We use here the former variant. The latter was used, e.g., in [9], where the embedded database MapDB[7] was used for this purpose.

3.2 Graph Pattern Tree of Variations

An important feature of our approach is that a new index can be created for each different graph pattern.

Due to labelling nodes and edges of GDB, patterns of the same structure can occur in different variations (see Fig. 2). All variations of a single graph pattern can be organized into a tree-like structure, called *graph pattern tree of variations*. A part of such tree for a triangle is shown in Fig. 2. Nodes represent individual graph pattern variations. A root node of the tree is reserved for the basic graph pattern variation with no additional information about nodes and relationships. Children of each node represent variations that provide some additional information compared to its parent nodes (i.e. when traversing deeper in the tree, more information about graph pattern is specified).

When querying a particular graph pattern in the database, one can use either a specific graph pattern variation, or arbitrary ancestor graph pattern from the pattern variation tree.

The more the database engine knows about the queried graph pattern, the more it filters graph space based on such specific information. Especially, node and edge labels which are usually already indexed by value-based indexing methods can be used.

That means if one wants to query the basic graph pattern variation that represents the root node of its graph pattern tree of variations, it will be faster to query it using the structure-based index proposed in this paper. On the other hand, querying variations that already provide a lot of information about nodes and relationships (such variations are situated on the lowest levels of the tree, including its leaves) it may be faster to use a value-based indexing method.

In other words, there is a level of the tree, where querying variations on such level stops being more effective when using the structure-based index compared to value-based index. The height and the depth of the tree depends on the graph pattern the index was created for and data within the database.

---

[7] http://www.mapdb.org/, last accessed 2018/11/14.

3.3 Updating Graph Pattern Index

A graph pattern index maps all graph pattern units that are matched by a graph pattern the index was created for. Such graph pattern units exist within the database and so can be manipulated via DML operations. Thus, they can be updated in such way they no longer match the graph pattern. Also, when adding new data to the database, new graph pattern units can emerge. For that reason, each graph pattern index must always map its graph pattern units that currently exist within the database. That means each index must be updated each time a DML operation is applied on the database. Otherwise, indexes would not provide reliable information when queried.

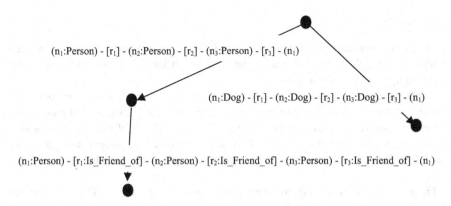

Fig. 2. Tree-like structure with graph pattern variations [8].

The intended DML includes operations like creating a node, creating a relationship, deleting a node, deleting a relationship, updating a node, and updating a relationship. Except the first one, all these operations affect the index, i.e. the index must be updated. It is done so within the same transaction that executed a DML operation. If a transaction is successfully committed, indexes will be updated. If a transaction is rollbacked, indexes will remain in the same state as before the transaction was initialized.

4 Implementation

The method for indexing graph patterns, including operations to create an index, query using an index and update an index, is implemented for the Neo4j GDB engine.

The major benefit of Neo4j is its intuitive way of modelling and querying graph-shaped data. Internally, it stores edges as double linked lists. Properties are stored separately, referencing the nodes with corresponding properties.

It is not easy to describe our implementation on a limited space, because it is tightly bind to database engine itself (Neo4j version 2.2. was used). It uses and extends Neo4j internal classes in order to achieve better integration. On the other hand, the implementation is done in a conceptually clean way and its shorted description presented in this paper may also explain a lot about extension of a database engine by additional services (like indexing method) in general. Implementation details can be found in [14].

This section is organized as follows: in Sect. 4.1, the idea of structure-based index implementation inside the GDB is explained, it provides necessary terminology used in the rest of the section. Section 4.2 describes an architecture of Neo4j and GraphAware framework classes involved in our implementation and explains the concept of our implementation. Sections 4.3, 4.4 and 4.5 claim to explain implementation of individual usage of structure-based indexes, i.e., their creating, using for querying and keeping them actual with DML operations over the database. Section 4.6 provides information how our implementation can be installed and used.

4.1   Overview

Our index implementation is done in the same GDB as basic graph data. We introduced an additional graph representing all indexes in the database. This graph has a root providing approach to all indexes.

Implementation of an index consists of a two-level tree. The first level has one node representing the index and containing appropriate metadata. This top-level index node is related to common root mentioned above. The second level of index representation consists of a set of graph pattern units. Each unit represents one pattern (triangle in our case). There are direct relationships to appropriate nodes in the database from each pattern unit.

Figure 3 describes an implementation of triangle shape index. The first and the second layer in the figure represents pure index data, while the third layer represents original data in GDB.

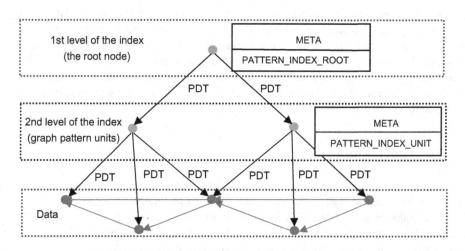

Fig. 3. Implementation of triangle shape index [8].

Index data is separate from original data on the logical level. This is done by assigning a special META label to nodes of the index and a special

PATTERN_INDEX_RELATION type (label) to relationships, they are denoted PDT in the figure. A special additional labels are used to differentiate the index root (PATTERN_INDEX_ROOT) and second level nodes representing particular indexes (PATTERN_INDEX_UNIT).

For the following discussion it is important to understand notions of graph pattern units and index units.

Graph Pattern Units. They are formed from actual data in the database. They are uniquely identified by relationships. In our example with triangle shape we may care about permutations in the result set or there may be more relevant relationships between two nodes, i.e. multigraph.

Index Units. Index units belong to index structure. They are uniquely identified by (ordered) list of its nodes. One index unit may point to several graph pattern units.

4.2   Architecture of Implementation

Implementation Background. Neo4j GDB engine is written in Java. It provides Core API which is used to communicate with a GDB. It is also used in our implementation.

GraphDatabaseService interface, the key part of Core API, is used as the main access point to a running Neo4j instance. It provides many methods for querying and updating data, including operations to create nodes, get nodes by id, traverse a graph and many more. Each of such operations must be executed within a transaction to ensure ACID properties. More than one operation can be applied to a database within a single transaction. For this purpose the interface also provides a method to create a new transaction. Neo4j then enables to use Transaction interface to build transactions in a very easy way. When using this interface, all operations within a transaction are enclosed in try-catch block.

Node and Relationship are other two important interfaces provided by Core API. Node interface provides methods that cover all possible operations with a single node, including manipulation with its properties, labels, and relationships. Relationship interface, on the other hand, provides methods to mostly accessed information about such relationship, including its end nodes or type. Note that it is not possible to change a type of a relationship in Neo4j. Such operation must be simulated by deleting and re-creating a relationship. It is also important to mention that Neo4j supports only primitive data types when storing properties of nodes and relationships. Thus more complex data types must be converted to string values before storing within their properties.

Since Neo4j 2.2, GraphDatabaseService interface enables to execute queries using Cypher. For this purpose a method execute is provided. Cypher query is passed as a string parameter to this method. Result of such query is organized in a table and returned as an instance of Result class. It is not necessary to enclose such operation in transaction try-catch block since it is, by default, executed within a transaction.

Graph Index Implementation. The implementation of the method of indexing graph patterns involves several classes. `PatternIndexModel` class is the core class of the method.

Figure 4 shows classes and their relationships. A single instance of `PatternIndexModel` class is created for a Neo4j database. It handles all operations that involve manipulation with graph pattern indexes. `PatternIndexModel` uses a singleton pattern, thus its instance can be obtained by calling `getInstance` method of the class. Instance of Neo4j database is passed as a parameter to this method. The class further provides methods to create a new index, query using an existing index, delete an existing index, and to handle updating of all existing indexes at once.

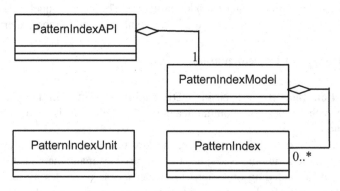

Fig. 4. Main classes.

Instance of `PatternIndexModel` manages all indexes that are created for a database. Each of these indexes is represented as an instance of `PatternIndex` class. Such instance holds basic information about an index including its name, graph pattern that the index was created for and other information that is necessary for its functionality. It also holds a root node of its index tree structure. All existing indexes are accessible within the instance of `PatternIndexModel` class via a single map. Such map consists of key-value pairs, such that a key represents a name of an index and a value represents appropriate instance of `PatternIndex`.

As mentioned above, each index tree consists of a single root node and index units that map actual graph pattern units within a database. Instances of `Pattern IndexUnit` class are used to represent specific index units. An index unit is identified by a group of nodes that form graph pattern units mapped by such index unit. Each index unit can map one or more graph pattern units. Thus each instance of `PatternIndexUnit` holds a set of nodes that identify it and a set of string identifiers that represent specific graph pattern units that belong to it. Identifiers of graph pattern units are persisted within a `specificUnits` property of appropriate index units in a database.

In order to query the database using index, the abstract class `QueryParse` is created. Two classes named `PatternQuery` and `CypherQuery` implement this class and are used for querying itself.

Additionally the class `DatabaseHandler` which provides manipulating methods for creating and processing individual parts of the index is provided.

4.3   Index Creation

To create a new index, `buildNewIndex` method is provided within `Pattern IndexModel` class. When calling such method, a user must provide a name of the index and also a graph pattern that the index should be created for. Cypher, more specifically its MATCH clause, is used to express such graph pattern. First of all, the process of graph pattern validation is applied to given graph pattern.

If the graph pattern is valid, all graph pattern units that match given graph pattern are found within a database. This is done by executing a simple Cypher query. The process of matching graph pattern units using a simple Cypher query is very time consuming. From each node within a database depth-first search or breath-first search (it is based on Neo4j settings) is applied in order to find matches for a graph pattern starting from such node. By applying such process a group of matched graph pattern units is retrieved. Some of them might be mutually automorphic. This is caused because of the fact that searching is done from each node individually.

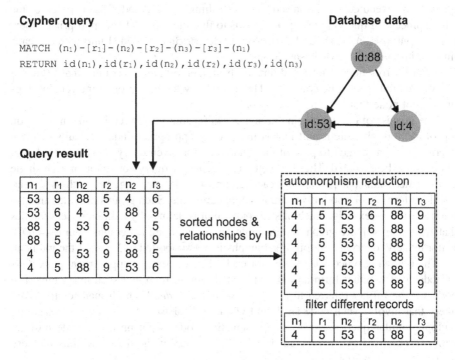

Fig. 5. Creating index - automorphism reduction.

Then such group can be also referred to as a group of automorphism groups of graph pattern units. Figure 5 shows a result for a query. In this case, the database consists only of a single triangle. There should be a single triangle matched but instead there are six records (i.e. graph pattern units represented in rows) in the result. Note that all of them belong to a single automorphism group. It is necessary to reduce automorphism in a group of matched graph pattern units such that each automorphism group consists of only a single graph pattern unit. For this purpose records are sorted by IDs of nodes and IDs of relationships separately and then only different records (i.e. graph patterns) are filtered.

After a group of graph pattern units is found within a database and automorphism is reduced, an index tree structure is built. Its index units will map these graph pattern units. Identifiers are created for specific graph pattern units and then stored within specificUnits property of appropriate index units. Also basic information about an index, including its name and graph pattern with parsed identifiers, is stored within properties of its root node. After that, a new index is stored under its name within patternIndexes attribute of PatternIndexModel instance.

4.4 Index Querying

To query using an index, getResultFromIndex method is provided within PatternIndexModel class. A user must provide an index to be used and also a query to be executed within an index when calling this method. The first step of the whole process of querying using an index is to find the root node of appropriate index. For this purpose an instance of PatternIndexModel is loaded into memory each time an instance of Neo4j is started up.

After the root node is retrieved directly from memory, all index units that belong to appropriate index must be collected. This is done by traversing outgoing relationships of the root node that head to these units.

The process of querying using an index can be split into executing given query on top of each graph pattern unit that is mapped by appropriate index. Results of these queries are then merged to present the final result for given query. That is indeed a lot of queries to be executed. Thus it is better to execute the query not on top of each single graph pattern unit but perhaps on each group of nodes that together with their relationships form one or possibly more graph pattern units. Such groups of nodes are also referred to as index units. Each of such index unit is defined by a different set of nodes that form one or more graph pattern units. Index units are subgraphs within a GDB. Unfortunately, querying on top of subgraphs is not supported in Neo4j at the moment.

The following process is applied for each of index units of appropriate index. One of nodes that together form an index unit is chosen to represent such index unit. The node will be further referred to as a *representative node*. Then all matches for given query that involve such node are found within a database. This is done by executing multiple similar queries, where a representative node is put on every node position within a graph pattern expressed in a MATCH clause of the query. Results of these queries are then merged.

In Cypher, this can be solved by using a single composed query. Figure 6 shows such composed query for a single index unit, which is represented by a node with ID 53.

In general, a single composed query consists of s subqueries, where s is the number of nodes within a graph pattern expressed in a MATCH clause of given query. Note that a graph pattern for which appropriate index is created consists of the same amount of nodes. As said at the beginning, this whole process is done for each index unit of appropriate index. Thus it is necessary to execute n such composed queries, where n is the number of index units of appropriate index. Finally, results of these queries are merged to present the final result for given query. This approach is used when implementing the method of indexing graph patterns.

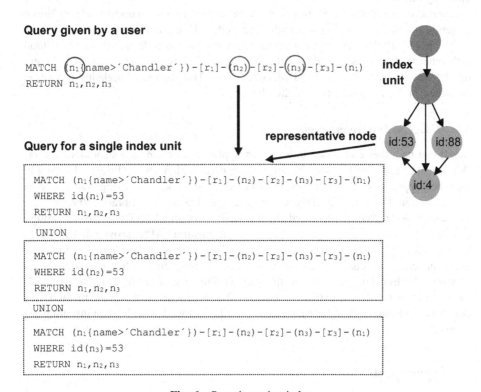

Fig. 6. Querying using index.

4.5   Index Update

Indexes are updated automatically each time a DML operation is applied to a database. For this purpose handleDML method is provided within PatternIndexModel class. This method is called inside beforeCommit method provided within GraphAware framework. Such method is called each time a transaction is about to be executed. It also provides its users with an instance of ImprovedTransactionData that holds all nodes and relationships that are being created, deleted, and updated by currently running transaction. Such instance is passed to the handleDML method as a parameter.

In the context of `beforeCommit` method it is possible to traverse a graph in the state as it was before a transaction was executed. Let's say a user deletes a relationship within a transaction. Then `beforeCommit` method is called. Instance of `ImprovedTransactionData` holds the relationship within a list of deleted relationships. At this moment, one can, e.g., explore end nodes of such deleted relationship. When executing Cypher queries using execute method in this context, a database is always in the state as it would be if the transaction was successfully committed. These are key functionalities used when implementing methods for updating indexes.

`handleDML` method performs some additional operations to update indexes before currently running transaction is actually committed. If the transaction fails and it is to be rollbacked, all these additional operations are reverted as well. `handleDML` method calls other methods based on data that is being modified by the transaction. All of them update indexes based on specific DML operations. The process of updating indexes is described in the analytical part of this thesis.

4.6   Usage of Implementation

Neo4j can be either embedded in a custom application or run as a standalone server. When using embedded mode, Neo4j can be used within an application by simply including appropriate Java libraries. In this case, the method of indexing graph patterns can be used in the same way. Thus it can be included as an extra library. When using a standalone mode, Neo4j is accessed either directly through a REST interface or through a language-specific driver. For this case, a custom API, provided by `PatternIndexAPI` class, is exposed to access all operations of the method of indexing graph patterns. In this case, the method must be built first. After it is built, it is necessary to drop built jar file into the plugins directory of appropriate Neo4j installation. In other words, the method is used as a Neo4j plugin. Note that the method uses GraphAware framework libraries, so these must be included as well when installing the method.

5   Experiments

Experiments were done on three different graph datasets: Social graph with a triangle index, Music database with a funnel index, and Transaction database with a rhombus index. These datasets are shortly described in Sect. 5.1. In Sect. 5.2, we present experiments done with growing the database size and the structure index effectivity. Measurements are done on social graph with triangle index. Section 5.3 presents measurements on all three datasets. Here only one size of each dataset is used, but we present also time necessary to keep index actual along with DML operations on the database. Some hypothesis about the size of index are provided in Sect. 5.4. Section 5.5 brings a discussion of the measurements.

All results are achieved by measuring within a test environment provided by GraphAware framework. The following configuration is used when performing measurements: 2.5 GHz dual-core Intel Core i5, 8 GB 1600 MHz memory DDR3, Intel Iris 1024 MB, 256 GB SSD, OS X 10.9.4.

To achieve the most accurate results, measurements are always performed multiple times and their results are averaged. Measuring is done for all cache types provided by Neo4j, i.e., no-cache (Neo4j instance with no caching), low-level cache, and high-level cache.

5.1   Graph Datasets

Social graph is a database that contains information about people and friendships between them. People, represented by nodes, have names and are distinguished to males and females by appropriate labels. Friendships between them are represented by relationships of Is_friend-of type. Such database of changeable size is generated by Erdős– Rényi model for generating random graphs (see, e.g., [2]). The generator is a part of used GraphAware framework[8]. Triangle index is built for a triangle graph pattern expressed in Cypher in Sect. 3.1.

Music database stores data about artists, detailed information about the tracks they recorded and labels that released these records. The database has a fixed size of 12 000 nodes and 50 000 relationships. It is one of the example datasets that Neo4j provides on its website[9]. The database contains 86 funnel patterns. Funnel index pattern (see, Fig. 7a) we used for this database has the following Cypher expression:

$$(n1) - [r1] - (n2) - [r2] - (n3) - [r3] - (n1) - [r4] - (n4)$$

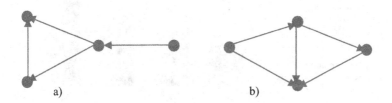

Fig. 7. Graph patterns used for Music DB (a) and Transaction DB (b) [8].

Transaction database stores data about transactions between bank accounts in a simplified way. Bank accounts, represented by nodes, are identified by account numbers. Transactions between bank accounts are represented by relationships. They have no properties on them since it is not crucial for the measurements. If used in a real database, they would probably hold some specific characteristics about them, e.g., a date of transaction execution or the amount of transferred money within a transaction.

Such database of changeable size was generated by a Cypher script that was created especially for this purpose. Such simple script creates bank accounts at first and then

---

[8] https://github.com/graphaware/neo4j-framework, last accessed 2018/11/14.

[9] http://neo4j.com/developer/example-data/, last accessed 2018/11/14.

generates a transaction relationship for each pair of these accounts with a given probability. The database we generated has 10 000 of nodes and 100 000 of relationships, it contains 70 rhombus patterns which were indexed. Rhombus index (see, Fig. 7b) is used for this database. It is formulated by the following Cypher expression

$$(n_1) - [r_1] - (n_2) - [r_2] - (n_3) - [r_3] - (n_1) - [r_4] - (n_4) - [r_5] - (n_2)$$

5.2 Measurement on Social Graph Database with Triangle Index

The size of the database scales from 50 nodes and 100 relationships to 100 000 nodes and 500 000 relationships. A matching triangle graph pattern using a simple query (i.e. without a graph structure index) is nearly impossible for larger databases of this type.

There are two metrics used: time and the number of database hits (DBHits), i.e., total number of single operations within Neo4j storage engine that do some work such as retrieving or updating data.

DBHits metrics for varying size of databases are shown on Fig. 8. We can see that from the database of size 10 000 nodes/50 000 relationships index pattern is much more effective than "simple query" (i.e. the query without index usage). The amount of DBHits for index grows linearly with growing the database while it grows exponentially for simple query.

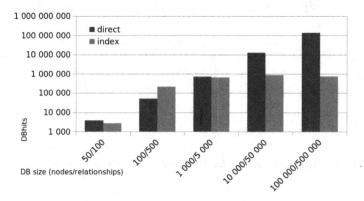

Fig. 8. Social graph, triangle index – DBhits metrics [8].

Time metrics are shown on Fig. 9. Again, we can see an exponential growth of time for a simple query and a linear growth for index. For the largest database of 100 000 nodes and 500 000 relationships, a query using an index is approximately 170 times faster and performs approximately 180 times less database operations than a simple query.

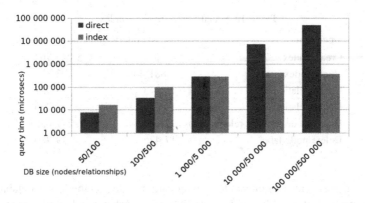

Fig. 9. Social graph, triangle index – time metrics [8].

5.3 DML Operations and Queries Measurement

The index must be updated together with DML operations on the base data. In our implementation update of the index is done in the same transaction as DML statement. We did measurements on all three databases mentioned in Sect. 4.1 for the following DML operations:

- creating an index,
- creating a relationship,
- deleting a relationship,
- deleting a node, and
- deleting a label of a node.

All these DML operations may affect existing indexes.

All measurements were done again for all cache types provided by Neo4j, i.e.:

- without caching,
- low level cache (i.e. file buffer cache), and
- high level cache (object cache).

The last mode is the most suitable for our purpose and, not surprisingly, it provides the best performance. For measurement results using another cache modes see [14].

In the Tables 1, 2 and 3 we present also a time of given operation without index usage to show additional costs for index maintenance. In Table 1, we present measured values done on a social graph with triangle index on the database having 10000 nodes and 50 000 relationships. Let us note, that there were 183 graph patterns on 179 nodes.

Table 1. Social graph, triangle index.

Operation	Simple query [μs]	Index [μs]
Create index	–	5 242 762
Query index	6 881 440	403 375
Create relationship	25 973	29 987
Delete relationship	146 820	157 726
Delete node with its relationships	228 399	277 835
Delete node label	17 475	19 380

Table 2 presents measurements for funnel pattern index on the Music database. The database consists of 12 000 nodes, 50 000 relationships and contains 86 funnel patterns.

Table 2. Music database, funnel index [8].

Operation	Simple query [μs]	Index [μs]
Create index	–	11 810 818
Query index	6 077 701	162 175
Create relationship	148 194	162 175
Delete relationship	289 220	283 263
Delete node with its relationships	484 171	665 391
Delete node label	64 420	66 610

Table 3 presents measurements for rhombus pattern index on the Transaction database. Database consists of 10 000 of nodes, 100 000 of relationships and contains 70 rhombus patterns.

Table 3. Transaction database, rhombus index [8].

Operation	Simple query [μs]	Index [μs]
Create index	–	36 238 883
Query index	41 794 378	1 243 503
Create relationship	29 659	64 432
Delete relationship	257 094	283 067
Delete node with its relationships	375 308	459 808
Delete node label	17 485	22 420

5.4   Index Size

Index size linearly grows with the number of pattern units found in the database and it also linearly grows with the number of nodes that the indexed pattern consists of. The index size can by asymptotically expressed as

$$\Theta(n_u * n_n)$$

where n_u represents the number of pattern units found in the database and n_n represents the number of nodes that the indexed pattern consists of. The exact number of nodes needed for the index is

$$n_{nodes} = 1 + n_u$$

where a single node represents the root of the index and n_u nodes represent individual pattern units found in the database. The exact number of relationships needed for the index is

$$n_{rels} = n_u + n_u * n_n$$

where n_u relationships connect the root node with all n_u pattern unit nodes. n_n relationships then connect individual data nodes belonging to a single pattern unit to its representative pattern unit node.

Table 4 presents index size using 3 different patterns and databases. There are 183 pattern units indexed in the first database which is more than double what is indexed in other two databases. Triangle pattern consists of 3 nodes whereas funnel and rhombus patterns consist of 4 nodes. This results in approximately the same size (in Mb) of index for all of 3 measured databases and patterns.

Table 4. Database and index sizes.

Database index	DB size (Mb)	Index size (Mb)	Pattern units
Social graph, triangle index	19,6	0,15	183
Music database, funnel index	89,6	0,2	86
Transaction database, rhombus index	17,2	0,1	70

5.5   Discussion

Let us note several interesting observations coming from our measurements:

- index creation time is not much higher than query time without an index for triangle and rhombus patterns, it is nearly two times higher for funnel index,
- query using an index is faster than the query without index for all three patterns, it is 17 times faster for triangle index, 112 times for funnel index, and 33 times for rhombus index,

- time to update the index in case of insert/delete a node or a relationship is on average 17% of time needed for DML operation itself.

Let us generalize the measurement and state several hypotheses about the efficiency of our implementation of pattern indexes:

- It was shown that starting from databases of size 10 000 of nodes and 50 000 of relationships queries using pattern indexes are more efficient than queries without them.
- Efficiency of pattern index increases with growing the database. Time and amount of database operations grow linearly for (triangle) pattern (Figs. 8 and 9).
- Complexity and size of the pattern used for index influence characteristics and efficiency of an index. We tested triangle, funnel, and rhombus patterns – all tested indexes are more than 17 times faster for querying, this ratio will growth with the size of the database.
- Time for keeping indexes actual seems to be under 20% of time necessary for DML operation.
- Experiments were done on the most general graph pattern indexes (the root of a graph pattern tree of variations, see Sect. 3.2 and index shapes in Figs. 1 and 7 for details).
- For practice usage, it is necessary to investigate the optimizer module for GDB engine. The module is responsible for creating execution plans including the decision when and which index is used for particular query evaluation.
- Method proposed in this paper stores data of structure-based indexes along the data which are indexed in the same database. It mixes together data and metadata and may cause undesirable effect on evaluation of queries which are not using structure based index.
- Unfortunately, Neo4j (and also others GDBMS) do not provide a separation space for data and metadata as it usual in relational database management systems.
- We already investigated another approach: store data about structural-based indexes in a separate storage, see [7] for details. Results we achieved are very similar to that presented in this paper.

6   Conclusions

In the paper a new method for indexing graph patterns was analyzed, designed and implemented for Neo4j GDBMS in order to speed up the process of matching graph patterns. The method enables to create, use and update multiple indexes, each created for a different graph pattern. Index data are organized in a tree structure and they are stored within the same database as the base data. This solution provides really fast approach from the index structure to data. On the other hand, it mixes index data and base data together in one common storage. It may negatively affect the evaluation of queries that do not use index patterns. We plan to address this issue in following research. It is the part of a more general topic how to store metadata and separate them from base data in GDBMS.

It is proved that using indexes which are created by the method introduced in this paper is beneficial for the process of matching graph patterns. In some cases queries using such indexes are extremely faster than simple Cypher queries. The paper aims to introduce the topic of indexing graph patterns and provides one of possible ways how to speed up the process of matching graph patterns within a GDB.

Acknowledgments. This work was supported by the Charles University project Q48.

References

1. Aggarwal, C.C., Wang, H.: Managing and Mining Graph Data. Springer, Boston (2010). https://doi.org/10.1007/978-1-4419-6045-0
2. Goldenberg, A., Zheng, A.X., Fienberg, S.E., Airoldi, E.M.: A survey of statistical network models. Found. Trends Mach. Learn. **2**(2), 129–233 (2009)
3. Mpinda, S.A.T., Ferreira, L.C., Ribeiro, M.X., Santos, M.T.P.: Evaluation of graph databases performance through indexing techniques. Int. J. Artif. Intell. Appl. (IJAIA) **6**(5), 87–98 (2015)
4. O'Neil, P.E.: The SB-tree: an index-sequential structure for high-performance sequential access. Informatica **29**, 241–265 (1992)
5. Pokorný, J.: Graph databases: their power and limitations. In: Saeed, K., Homenda, W. (eds.) CISIM 2015. LNCS, vol. 9339, pp. 58–69. Springer, Cham (2015). https://doi.org/10.1007/978-3-319-24369-6_5
6. Pokorny, J., Snášel, V.: Big graph storage, processing and visualization. In: Pitas, I. (ed.) Graph-Based Social Media Analysis, Chap. 12, pp. 391–416. Chapman and Hall/CRC, Boca Raton (2016)
7. Pokorný, J., Valenta, M., Ramba, J.: Graph patterns indexes: their storage and retrieval. In: Proceedings of the 19th International Conference on Information Integration and Web-Based Applications and Services (iiWAS 2018), Yogykarta, Indonesia, November 2018, 5 pages (2018)
8. Pokorný, J., Valenta, M., Troup, M.: Indexing patterns in graph databases. In: Proceedings of the DATA 2018, pp. 313–321 (2018)
9. Ramba, J.: Indexing graph structures in graph database machine Neo4j II. Master's thesis, Faculty of Information Technology, Czech Technical University in Prague (2015). (in Czech)
10. Robinson, I., Webber, J., Eifrém, E.: Graph Databases. O'Reilly Media, Menlo Park (2013)
11. Sakr, S., Al-Naymat, G.: Graph indexing and querying: a review. Int. J. Web Inf. Syst. **6**(2), 101–120 (2010)
12. Srinivasa, S.: Data, storage and index models for graph databases. In: Sakr, S., Pardede, E. (eds.) Graph Data Management: Techniques and Applications, Chap. 3, pp. 47–70. IGI Global, Hershey (2012)
13. Tivari, S.: Professional NoSQL. Wiley/Wrox, Hoboken (2015)
14. Troup, M.: Indexing of patterns in graph DB engine Neo4j I. Master's thesis, Faculty of Information Technology, Czech Technical University in Prague (2015). https://dspace.cvut.cz/bitstream/handle/10467/65061/F8-DP-2015-Troup-Martin-thesis.pdf?sequence=1&isAllowed=y
15. Ullmann, J.R.: An algorithm for subgraph isomorphism. J. ACM **23**(1), 31–42 (1976)
16. Yan, X., Yu, P.S., Han, J.: Graph indexing: a frequent structure-based approach. In: Proceedings of SIGMOD Conference, pp. 335–346. ACM (2004)

17. Yan, X., Han, J.: Graph indexing. In: Aggarwal, C.C., Wang, H. (eds.) Managing and Mining Graph Data. Advances in Database Systems, vol. 40. Springer, Boston (2010). https://doi.org/10.1007/978-1-4419-6045-0_5
18. Yuan, D., Mitra, P.: Lindex: a lattice-based index for graph databases. VLDB J. **22**, 229–252 (2013)
19. Zhao, P., Han, J.: On graph query optimization in large networks. VLDB Endow. **3**(1–2), 340–351 (2010)
20. Zhu, L., Ng, W.K., Cheng, J.: Structure and attribute index for approximate graph matching in large graphs. Inf. Syst. **36**(6), 958–972 (2011)

Architectural Considerations for a Data Access Marketplace Based upon API Management

Uwe Hohenstein[(⊠)], Sonja Zillner[(⊠)], and Andreas Biesdorf[(⊠)]

Corporate Technology, Siemens AG, Otto-Hahn-Ring 6,
81730 Munich, Germany
{Uwe.Hohenstein,Sonja.Zillner,
Andreas.Biesdorf}@siemens.com

Abstract. Data and access to data are becoming more and more a product or service to be sold. In order to deal with data access, this paper presents a marketplace for trading data. Within the marketplace, producers can present their offerings for data access and algorithms, whereas consumers are able to browse through the offerings, to subscribe to corresponding services, and to use them after being approved. As a specific property, the presented approach allows to control data access at a fine-granular level. This supports use cases where different consumers should see different portions of data according to subscribed price models and/or Service Level Agreements (SLAs). The approach takes benefit from API management, particularly the tool WSO2. The paper discusses possible architectures to combine API Management with user-specific filtering and control of SLAs, and illustrates in detail how the features of WSO2 help ease implementation and reduce effort on the one hand. On the other hand, the paper elaborates upon several technical challenges to overcome.

Keywords: Marketplace · Data access control · Filtering · API Management

1 Introduction

Data and related data-processing algorithms are increasingly becoming a good to sell, especially with the advent of integrated data analytics. Moreover, data and algorithms should form some kind of ecosystem in which exchange becomes possible in a protected and controlled usage. The ecosystem helps to bring together data providers and data consumers, which are different stakeholders. Indeed, public research on algorithms is often conducted outside a data provider's premises by research organizations that, e.g., bring in their expertise in machine learning. As a consequence, these organizations are mostly unable and not allowed to access the data. In fact, the data provider is not interested to give access to data for free. Unfortunately, promising value-adding algorithms and applications can only be tested if data is available. The other way round, the results of research activities and explored algorithms could give benefit to data owners. But since algorithmic research is expensive, profit-oriented research institutions want to earn money with algorithms and/or their results.

To get out of the dilemma, monetary incentives could help. This means access to data can be sold, especially to algorithm developers, while algorithms and the data they

© Springer Nature Switzerland AG 2019
C. Quix and J. Bernardino (Eds.): DATA 2018, CCIS 862, pp. 91–115, 2019.
https://doi.org/10.1007/978-3-030-26636-3_5

produce can be offered to data owners or other organizations the other way round. As an additional step, data can be made anonymous before granting access in case of regulations or data protection laws. In fact, such a trading platform constitutes some kind of marketplace for trading, sharing, and exchanging data, data processing algorithms as well as data analytics results. Certainly, the marketplace should be open for specific providers and consumers or the public. A marketplace enables providers to publish and propagate data and algorithms, thus hopefully obtaining attention by potential consumers. Data consumers in turn can browse through the offerings and, if interested, ask for usage. Security is obviously a mandatory prerequisite for such a marketplace. First, only granted customers must be allowed to use an offering, i.e., consumers have to authenticate successfully. Second, some filtering mechanism becomes necessary to give each customer access only to particular portions of data according to some rules. The portion might depend on an individual consumer, his/her organization, the selected Service Level Agreement (SLA), or any other specification. That is, filtering of data must be somehow consumer-specific, in a controllable and configurable manner, to give different consumers access to different parts of the data.

Research in the domain of databases addresses the problem of consumer-specific access to data sources (cf. Sect. 2). Several concepts have been developed for row-level access control (to filter out rows) and column-level access control (to mask out or omit specific columns) in relational database systems.

But as far as we noticed, only little attention has been paid to monetization of data and data access. Marketplaces can provide a potential platform to trade data access, but appropriate solutions are missing to integrate advanced database filtering. Moreover, existing filtering approaches from the database community turn out to be unable to handle quite a dynamic set of unknown users. In particular, a high degree of automation becomes indispensable. And finally, In order to be also able to limit data quality, the result size, or to throttle too many accesses depending on a chosen SLA, advanced concepts are required.

To tackle these challenges, we use API Management to implement a marketplace architecture. The marketplace forms a platform that allows data providers to publish and sell access to APIs. Figure 1 shows the providers' view of a system that we developed and tested in a funded project named "Klinische Datenintelligenz" [20]. The view shows all the offerings and allows for adding new and modifying existing ones. Consumers are enabled to browse through the published services in the marketplace in a similar manner. If being interested, they can subscribe and purchase access to a particular API. A consumer web interface, shown in Fig. 2, is offered for performing those tasks. High flexibility is achieved by combining API Management and flexible filtering. We illustrate that using API Management is beneficial as it already provides much functionality for implementing a marketplace. Indeed, API Management takes care of scalability for a high amount of (mostly unknown) users, OAuth-based authentication, and features for auditing, monitoring, and integrating billing concepts.

A previous paper [11] has already discussed the overall architecture. In this paper, we extend this paper by elaborating on some details. In particular, we handle specific points that are important to secure a marketplace. Thus, concepts are discussed how to avoid SQL injection and denial of service. Another concept is to throttle particular users if they consume too much resources or endanger the overall system.

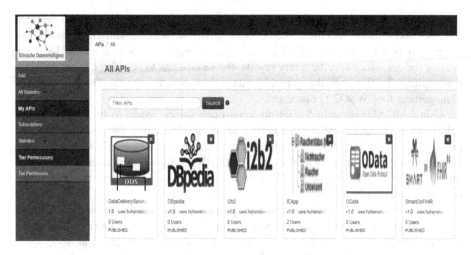

Fig. 1. Providers' web interface [11].

Fig. 2. Consumers' web interface [11].

In the following, Sect. 2 summarizes the related work as far as relevant for the paper. From the discussion, we deduce the necessity for our research.

In Sect. 3, we introduce the concepts of API Management. We explain why API Management reduces effort to implement a marketplace by already offering fundamental concepts. We particularly focus on the concrete WSO2 API Management tool.

Afterwards, a customizable Data Delivery Service (DDS), which provides features for row- and column-specific filtering, is proposed in Sect. 4. The DDS provides a generic query interface in a REST-based style. In order to enable the Data Delivery Service for consumer-specific query results, the service has to be offered as an API in a marketplace. To this end, the general strategy how to integrate the DDS in the API Management tool WSO2 is discussed.

Using API-Management for setting up a marketplace seems to be appealing, but unfortunately several challenges are still to tackle. Section 5 discusses the overall problem space in this context, especially the major challenges of combining API Management and user-specific filtering. Especially, a highly automated environment turns out to become necessary in the presence of an unknown and typically large consumer community.

Section 6 describes an architectural proposal with potential solutions and alternatives to tackle those challenges in an appropriate manner. Some more details about critical concepts such as SQL injection and denial of service are given in Sect. 7.

Finally, we conclude our work in Sect. 8 and discuss some future work.

2   Related Work

A lot of work has been done on various facets of access control restricting activities of legitimate users. [10] presents a general survey about research in this area.

Bertino et al. already addressed in 1999 the need for flexible access control policies [4]. Their well-defined authorization model supports a range of discretionary access policies for relational databases.

Since then, database vendors have also recognized that basic database mechanisms such as views, stored procedures, and application-based security are insufficient to provide an effective access control [17]. In the meantime, popular relational database products have caught up by introducing advanced security features. Despite being named differently, the concepts in the products are quite similar and allow for row-level and column-level access control. IBM DB2 offers row permissions and column masks definitions [17]. Column masks perform some customizable obfuscation of data by patterns for XXX-ing parts of phone numbers, accounts etc. Corresponding Microsoft SQL Server concepts are named Dynamic Data Masking, Column Level Permissions, and Row Level Security, whereas Oracle leveraged the Virtual Private Database technology [14]. The policies usually rely on stored procedures to be defined for each table. Moreover, the features require that each user has obtained an individual connect account, leading to increased administration effort. Opposed to our proposal, they achieve less flexibility, e.g., as no further SLAs can be integrated.

Upon those basic concepts, Barker [3] defines a high-level specification language for representing policies. The language is a formally well-defined, dynamic fine-grained tuple calculus for meta-level access control. Using query rewriting, any specification is translated into SQL in order to let the policy become effective. The problem of view proliferation becomes easier to manage than for user-based views like [16] by categorizing users to trust level, job level etc. Hierarchical and negative authorizations are also possible by means of policy withdrawal and overriding.

Another technique to control the disclosing data process has been discussed by LeFevre [13]. A high-level specification language allows for representing policy requirements for Hippocratic databases. A policy can be defined in P3P and EPAL and enables a subject to control who should be permitted to see his/her data for what purpose. Associating operations with a purpose and a recipient results in a view which obfuscates the values of particular cells with null, and diminishes rows from the query result that are protected by purpose-recipient constraints.

Rizvi [16] introduces a novel Non-Truman mode. While the Truman mode behaves similar to row/column level security, the Non-Truman mode tackles the disadvantage of possible misinterpretations of query answers because of suppressed records by row-level filtering. The approach relies on special authorization views for filtering. If a user query cannot be answered by using the authorization views only, the query is rejected. Otherwise, the query is valid and executed against the table without any modification.

Pereira et al. distinguish in [15] three general architectural solutions:

(a) Centralized architectures mainly applying views and parameterized views [18], query rewriting techniques [3, 16, 21], and extensions to SQL [6, 7];
(b) distributed architectures [5];
(c) mixed forms of architectures [8, 23].

They present a proposal that belongs to category (c) and focus on role-based access control to supervise an indirect access mode. In an indirect mode, SQL queries are executed in JDBC, Hibernate, LINQ etc. yielding some kind of result set, whereupon changes are performed and then committed to the database.

However, all these proposals neglect the marketplace aspects and the resulting challenges that we address. For instance, the approaches rely on individual database accounts for users, i.e., a shared connect is not possible and administrative effort is accelerated. Moreover, the approaches are unable to take SLAs beyond row/column level security into account.

Further approaches secure access control by generating application code. They operate at compile time and thus are not applicable to our work. Validating policies at compile time, programmers have to know database schemas and security policies while writing source code. In this sense, [1] presented a complete framework to define security aspects early in the software development process. Based upon a model, access control policies are derived and applied at compile-time. Similarly, the Ur/Web tool [7] allows programmers to write statically checkable access control policies. Programs are then inferred in such a way that query results respect a set of specified policies. In the SELINKS programming language [8], programs are written in a LINQ-like language called LINKS. The approach targets at building secure 3-tier web applications. A compiler produces byte-code for each tier to satisfy policies. Sensitive data does not become directly accessible by by-passing policy enforcement. [22] presents an approach to integrate access control to methods of remote objects via Java RMI. To this end, Java annotations that represent access control policies are attached to remote objects and methods. RMI Proxy Objects are generated in such a way that policies are satisfied. The proposal in [9] also uses annotations to attach a more fine-grained access control to methods. A novel tool to detect errors in role-based access policies is discussed in [12].

While various approaches have been discussed in the database area to limit visibility of data by filtering, only little attention has been paid to selling data access at marketplaces. The proDataMarket [19] approach focuses on monetizing real estate and related geospatial linked and contextual data. The paper suggests a marketplace architecture that offers a provider and consumer view similar to ours, however, the marketplace is not based upon API Management. Moreover, the paper does not handle the major challenges of user-specific filtering and handling SLAs.

3   API Management and WSO2

Wikipedia (https://en.wikipedia.org/wiki/API_management) defines API Management as "the process of creating and publishing Web APIs, enforcing their usage policies, controlling access, nurturing the subscriber community, collecting and analyzing usage statistics, and reporting on performance."

API-M provides a platform for producers and consumers handling their interplay. Figure 3 illustrates the technical infrastructure of the WSO2 API Management (API-M) tools. At the left side, providers can provision existing backend APIs to API-M. The backend API is defined by its URI and might be of REST or SOAP style. All offerings obtain a new URI by the API-M, which is internally mapped to the original backend service URI. Thereby, each provider a_provider obtains a base URI such as https://server:8243/t/a_provider. Every published API-M service extends the base URI by a service-specific part like /dds. Hence, API-M acts as a proxy for the backend service.

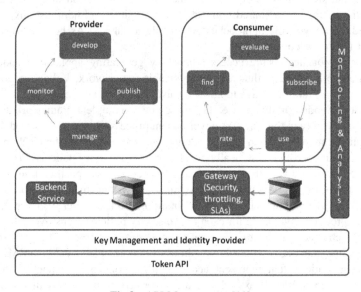

Fig.3. API Management [11].

API-M is a good basis for a marketplace for data-access applications, especially for a service delivering data. The approach we present relies on the WSO2 API Management: WSO2 is an open source API Management, which is also offered as a payable Cloud offering (the latter having special features already built-in). Figures 1 and 2 present some screenshots for a marketplace that we have set up for a funded project [20] in the medical area. Figure 1 illustrates the providers' web interface `https://server:8243/publisher`. The interface shows all the offered APIs as icons. Using the functionality at the left side, new APIs can be published by "Add".

Figure 2 shows the consumers' web interface `https://server:8243/store` with all the subscribed services (at the bottom), the possibility to sign-up to the API-M as a new consumer, to log in (both not visible in Fig. 2), and to subscribe to a service as a consumer.

A consumer can browse through all the offerings of all the providers (cf. left side of Fig. 2). If interested, a user can sign-up to the API-M with a name and password. Having then logged in, a consumer can subscribe to a particular service of his interest.

One important feature is the addition of an OAuth-based access control to even unprotected services. For using a subscribed API, a consumer has first to request an OAuth security token issued by the Key Management and Identity Provider. The token must be passed with every REST request in the "Authorization" header and is checked by the Gateway. The Gateway takes care of security, especially authentication, throttling etc. In particular, API-M checks the security token and allows invocation of the API by forwarding requests to the backend service only if the token is valid. The token can be requested by pushing the "Regenerate" button (cf., "Access Token" box in Fig. 2). The token expires after a configurable time, but can be renewed at any time.

We have chosen WSO2 as an implementation basis for our marketplace because it provides several useful features:

- Several dashboards are available in addition to the providers' and consumers' views. For example, a management view enables the administrator of the marketplace platform to administrate tenants, i.e., providers. Furthermore, an administration view gives each provider the possibility to manage his/her consumers.
- The provider of a service is able to approve every user sign-up and/or subscription in the administration view; corresponding pre-defined business processes can be triggered, and new ones can be specified. A self-approval without provider interaction can also be configured.
- APIs possess a life cycle, which are decoupled from the backend services. The life cycle is visible in the providers' view and can be controlled at runtime. An API can be either in the status of being *prepared* but not yet visible, deployed and published in the platform as a publicly usable *mock* (with the purpose to gather feedback about usability), *published* to become visible and open for subscriptions, *deprecated* for new subscriptions, or (maybe temporarily) *blocked* for all or particular tenants.
- An application's publication can specify information about usage conditions and pricing: To this end, APIs can be offered in several tiers such as Gold, Silver, or Bronze with certain SLAs and prices associated. With a subscription, a consumer accepts the usage conditions (e.g., about payment). Moreover, a throttling of access according to tiers can be configured without implementation effort.

- A billing component can be integrated, which is based upon consumption using a monitoring component that tracks all consumers' activities.
- There are powerful concepts to map a frontend URL as published in the API-M to the backend URL of the service, for instance, to change the URL by switching query parameters and path elements. Similarly, the request bodies and responses can be transformed.

```
{
  "columns": [  // columns of the result
    "encounter_num", "patient_num", "concept_cd",
    "provider_id", "start_date", "quantity_num",
    "units_cd"
  ],
  "types": [ // data types for those columns (same order)
      "int4", "int4", "varchar", "varchar", "timestamp",
      "numeric", "varchar"
    ],
  "elapsedMs": 36, // server-side execution (time in ms)
  "size": 2,    // number of records in result
  "content": [ // result records
    {
      "no": 0, // 1st record
      "values": ["1791", "1", ... ]
    },
    {
      "no": 1, // 2nd record
      "values": [ ... ]
    },
    ... // further records
  ]
}
```

Fig.4. DDS response.

4 Proposal for a Data Delivery Service (DDS)

4.1 A Generic REST API

Before diving into the details of the marketplace architecture, we first propose a Data Delivery Service (DDS) for relational database systems. The DDS offers a generic REST service that allows for executing arbitrary SQL queries passed as a string. The DDS is REST-based, but in fact, HTML/JavaScript based user interfaces can be put on top of the REST API. In fact, the DDS is supposed to be offered as a service in theWSO2-based marketplace. A DDS request has the following structure:

```
POST <Host>/<DBType>/<Database>/query?<Options>
```

<Host> is the host address of the server running the DDS, <Type> specifies the database product, e.g., PostgreSQL9.6 or Oracle13c, and <Database> refers to a particular database. <Options> can be used to control the outcome, e.g., by pagination. The connect information for all the database servers <DBType> is configured in a configuration file. The request body of the POST request contains the SQL query to be executed, for example:

```
SELECT encounter_num,patient_num,concept_cd,provider_id,
       start_date,quantity_num,units_cd,observation_blob
FROM   i2b2myproject.Observation_fact
WHERE observation_blob <> ''
```

This query refers to a medical i2b2 database. Setting the "Accept" header accordingly, the result can be requested in XML or JSON format. A JSON result looks as displayed in Fig. 4.

The result has a quite generic structure. The first two parts "columns" and "types" return the meta-data about the query result, i.e., the column names and their data types. The meta-data can be suppressed by applying the query parameter metainfo=none in <Options>. However, the meta-data is useful to let the result become interpretable and machine readable. "elapsedMs" yields the server-side execution time in ms, whereas "size" returns the total number of retrieved records. The last "content" part contains the records one by one according to the meta-data structure.

The DDS provides further features such as pagination with query parameters top= and limit=, streaming, and data compression during transport.

4.2   Deploying the DDS in WSO2

To integrate the DDS into WSO2, the providers' view can be used by choosing the "Add" button at the left side of Fig. 1. There are three steps to be performed: Design, Implement, and Manage. In each step, some, partially optional, specifications have to be done.

The first "Design" step defines the URI of the offered API and the operations. Name, Version and Context, have to be specified to altogether determine the tenant-specific URI of an API as <Server>/t/<Tenant>/<Name>/<Version>, for instance, https://server:8243/t/a_provider/dds/V1.0. A thumbnail icon for the API can be downloaded, e.g., for the DDS. Moreover, the visibility of the new API can be defined as "Public", "Visible to my domain" or "Restricted by roles"; the domain should be understood as the provider, e.g., siemens.com. An optional description can be attached to describe the functionality, the usage options, conditions, and SLAs. The offered HTTP verbs (GET, POST etc.) must be defined, a URI extension and as well as "Accept" and "Consumes" headers might be added. To group APIs to categories, tags are useful. For instance, a "Data Access" group can be set up, which then becomes visible in the consumers' web interface.

The next "Implement" step handles the connection to the backend service. The most important step is to specify the HTTP(S) Endpoint for the DDS Backend Service, for a production and possibly sandbox stage. Requests to the API Gateway are then forwarded to that endpoint. Moreover, a security scheme can be chosen: "Non Secured" determines that the backend service is not protected and does not require an authentication. If backend authentication is mandatory, credentials can be passed by using the "Secured" option.

Sometimes is it useful to transform the URI of the defined API to the backend URI and/or to modify request bodies and responses. To this end, a mediation for in- and/or out-going (In/Out Flow) requests is configurable: Here, "preserve_accept_header" and "apply_accept_header" are useful to pass "Accept" and "Content-Type" headers. A user-defined mediation is also possible, which can be specified in XML for simple mappings; more complicated mediations must be implemented in Java.

The final "Manage" step specifies further settings. First, the availability of tiers can be selected from the list of predefined tiers: standard tiers are Bronze, Silver, and Gold. Others can be defined by a tenant in the administrator dashboard (`https://localhost:9443/admin-dashboard/`, cf. Sect. 7.4). Next, subscriptions define which users of which tenant should be allowed to see the API: "Available to current tenant only", "Available to all the tenants", or "Available to specific tenants" are the options.

Having performed these three steps, a provider has exposed a DDS for his data to the API Management as a REST service. The new URI, e.g., `https://server:8243/t/a_provider/dds/V1.0` can be used by subscribed and acknowledged consumers. Every request is checked by WSO2 for correct credentials (by OAuth) and forwarded to the backend service URI thereby performing any specified mediation.

5 Challenges

Using API management (API-M) such as WSO2 for implementing a marketplace is helpful and useful because a lot of marketplace functionality like producer and consumer web interfaces is available. But API-M also introduces some challenges when offering a Data Delivery Service (DDS) as a service.

Any consumer interested in using the DDS can subscribe to the DDS service in the API-M. After a successful approval by the provider, a consumer can ask the API Management for issuing an OAuth security token; the OAuth token is consumer and DDS-specific. The security token has a specific expiration time and must be passed to the API-M as an integral part of any DDS invocation. API Management checks the validity of the security token and – if valid – invokes the Data Delivery Service.

Figure 3 shows that neither the DDS nor the database is directly accessed by consumers: API Management acts as a proxy in front and protects the DDS. API-M receives requests and forces consumers to authenticate with a valid OAuth token in order to let API-M forward the call to the DDS.

This is all functionality already offered by API Management. However, one important point is missing: We have to take care of consumer-specific filtering in the

DDS: Consumers should see specific data instead of the whole data set. Examples in case of relational database systems are row/column level filtering, reducing the size of a result set, or other SLA attributes that affect the quality of returned data. To achieve filtering, three major challenges have to be tackled.

- (C1) Information about the consumer is required for filtering. Any kind of filtering should take place depending on the consumer who is invoking the request. The problem is how to get the consumer's identity from the API-M, especially as the consumer uses a cryptical OAuth token for authentication at the API-M. Indeed, the DDS at the backend must know the consumer. Moreover, the selected tier can also be used to perform user-specific filtering (cf. Sect. 7.5).
- (C2) Suppose the consumer is somehow known by the DDS: The consumer information has to be used to reduce results in a flexible manner. Moreover, filtering should take place with as little manual administrative effort as possible. The software of the DDS as a backend service should not be modified for every new consumer.
- (C3) Finally, API-M and DDS have to co-operate. In particular, there is an unknown consumer base, with potentially many unexpected consumers. Each provider wants to give consumer-specific access to the database, but does not know about potential consumers of the marketplace. Moreover, each subscribed user requires certain actions to be performed in the database such as creating some database access roles.

6 Architectural Proposal

The following section tackles the challenges (C1) to (C3). The discussion relies on the API Management tool WSO2, however, other tools offer similar concepts.

6.1 Passing Authorization Tokens

As already mentioned, every invocation of an API via the API-M requires a security token that is issued by API-M for every subscribed consumer. This means that there is no obvious mean to pass the consumer information to the DDS besides the security token. However, the user information is required by the DDS in order to perform filtering. The assumption is that the security token somehow bears the consumer information. Concerning Challenge (C1), there are two major problems to solve: How to pass the security token to the DDS and how to interpret the quite cryptic token like 6db65cdf3231be61d9152485eef4633b in the DDS.

We investigated that WSO2 API-M can be configured to pass the security token (coming from the consumer) to the DDS backend service. There are mainly two steps to perform within the provisioning of an API. A predefined mediation con-figuration "preserve_accept_header" can be applied to the request handling (cf. Sect. 4.2). Then, WSO2 passes an x-jwt-assertion header to the backend service, i.e., DDS. This header contains a so-called Java Web Token (JWT) that possesses 1445 bytes and looks even more cryptic than the security token. However, such a JWT can be parsed with a common JWTParser like http://jwt.io. Even more important, the parsed JWT yields the information the DDS requires. Figure 5 shows the content that can be extracted from a JWT.

That is, the DDS is finally able to obtain and interpret the JWT and to extract information about `tier` or `enduser` from the token.

```
{
  "iss": "wso2.org/products/am",
  "exp": 1467298671690,
  "http://wso2.org/claims/applicationid": "1",
  "http://wso2.org/claims/applicationtier": "Unlimited",
  "http://wso2.org/claims/apicontext": "/dds/v0.1",
  "http://wso2.org/claims/version": "v0.1",
  "http://wso2.org/claims/tier": "Silver",
  "http://wso2.org/claims/keytype": "PRODUCTION",
  "http://wso2.org/claims/usertype": "APPLICATION",
  "http://wso2.org/claims/enduser": "userA@companyB.com",
  "http://wso2.org/claims/enduserTenantId": "-1234"
}
```

Fig.5. Decrypted token.

6.2   Options for Implementing Filtering

Having extracted and parsed the OAuth token, the consumer information becomes available to the DDS. Concerning challenges (C2) and (C3) in Sect. 5, there are two major alternatives.

(A) The extracted consumer can be used to connect to the database and to perform the filtering in the database by using the database features that are discussed in Sect. 2. This is an easy option, but has the drawback that for every new consumer (which is just known after subscription), a corresponding database user has to be created for database logon, and further GRANTs and database-specific actions are required. This must be integrated into the subscription approval business process. Hence, there is some administrative effort to keep API-M and database users in sync. As a consequence, self-approval of consumers is complicated. Moreover, a user and password management is required to secure the database. The security token might be a good candidate for the password. However, the token has a short expiration time in the area of some minutes; the user password that the consumer has chosen during sign-up for the API-M is invisible. Another disadvantage is the huge amount of database accounts. This leads to many parallel open database connections – one for each consumer – since connection pools usually do not work with a user-spanning pooling over database accounts. This has a major impact on the performance and raises high resource consumption in the database.

(B)   As an alternative, it is possible to use one single connect string to connect to the database. This requires less administration effort in the database and also solves the connection pooling issue. Unfortunately, database filtering features can no longer be used since the real consumer is now unknown in the database due to the shared database account. Hence, filtering must be performed in the DDS with some implementation effort as a consequence. As a gain, this offers most flexibility for filtering, especially since further information such as the chosen tier can be taken into account. There are two major options:

(B.1)   The Data Delivery Service can perform filtering by query rewriting and sending consumer-specific, modified queries to the database. There are many approaches in the literature, for example [3, 16, 21]. That is, the filtering logic becomes part of the DDS, and the DDS has to be aware of the corresponding policies. The DDS must know which user is allowed to see what columns and rows. Here, it is important that query rewriting at runtime is a must.

(B.2)   Alternatively, the DDS can submit the original query and perform the filtering on the received result sets. Again, the DDS has to know the filtering policies. Performing complex filtering results is challenging leading to some complex analysis, since the filter conditions must be interpreted; the columns are not directly obvious and must be derived from the submitted SQL query. Moreover, there are performance issues since such a client-side vertical filtering transports larger query result sets from the database to perform filtering.

6.3   Database Connect

We suggest a more generic approach combining the alternatives to achieve best benefit. In a nutshell, we proceed as follows:

1. In order to solve the issue with several database accounts, the DDS connects to the database with a *shared* account in the sense of alternative (B), without giving any further privileges to this account except for the permission to connect.
2. However, the consumer is passed as a hook-on to the connection. This allows us to set up specific user access privileges for the consumer in the database.
3. The consumer information is used to control the result for a specific consumer by one *single* database view for all the users. This eases the administrative effort. Sect. 6.4 will present the details.
4. Further filtering can occur in the DDS to achieve powerfulness, e.g., considering the tier and/or scope.

Indeed, we found for (1) and (2) some quite hidden, product-specific mechanisms that are available in some database systems and enable passing consumer information to the database. For example, in PostgreSQL it is possible to create an account to connect to the database. The account has no further access to tables and views beyond the allowance to login:

```
CREATE ROLE loginOnly NOINHERIT LOGIN PASSWORD 'pw';
```

The DDS uses this `loginOnly` role. Another account `a_user`, created without the `LOGIN PASSWORD` option, is unable to connect:

```
CREATE ROLE a_user;
```

However, if `a_user` is added to the `loginOnly` group by GRANT `a_user` TO `loginOnly`, DDS can login with `loginOnly` and issue the statement SET ROLE `a_user`. This lets the user privileges for `a_user` become effective for every successive query.

Teradata has a so-called Query Band mechanism that behaves similarly from a functional point of view. SQL Server offers an `EXECUTE AS` statement to switch the user after having connected.

Letting the DDS authenticate in the database with a (shared) predefined account avoids a complex user management and corresponding administrative effort as well as negative performance impact.

6.4   Column-and Row-Level Filtering

The goal to achieve is a flexible filtering approach with little administrative effort. We present a hand-made solution to become product-independent. However, note that the approach also allows for integrating with database features or the research approaches discussed in Sect. 2.

Our approach consists of *one* common view for each base table handling the filtering in a consumer-specific manner. Accessing the common view, consumer information is implicitly used to restrict results. The view performs column- and row-level filtering based upon the consumer in a generic manner.

The filtering principle uses additional tables similar to [3]. A configuration table `Privileges` (cf. Table 1) controls the visibility for users. For example, if `TabXColumn1` contains a value `false` for a particular consumer, then `Column1` of table `TabX` should not be visible for that user. Consequently, the configuration of user privileges – which columns and which rows in tables should be visible – is done external to the DDS.

Table 1. Privileges.

Privileges	Consumer	TabXColumn1	TabXColumn2	...
	User1	true	false	
	User2	false	true	
	User3	true	true	

This table is used in a generic view `TabX_Filter` to be created for each table `TabX`:

```
CREATE VIEW TabX_Filter
SELECT CASE WHEN  p.TabXColumn1
            THEN  t.Column1
            ELSE  NULL END AS Column1,
       CASE WHEN  p.TabXColumn2
            THEN  t.Column2
            ELSE  NULL END as Column2,
       ...
FROM TabX t
LEFT OUTER JOIN Privileges p ON p.consumer = current_user
AND  <Condition>  // explained later
```

Fig.6. Filtering view.

Certainly, each user is withdrawn access to the base tables TabX; only access to the TabX_Filter views is granted.

CASE expressions nullify or mask out columns for dedicated users according to what is defined in table Privileges. Hence, the behavior is similar to the nullification semantics of [13].

The views need the current user (as being hooked to the connect). Database products usually provide corresponding functions, e.g., there is the current_user function in PostgreSQL. Oracle offers a so-called system context that is accessible in a similar manner.

Row-level filtering, i.e., <Condition> in the view definition of Fig. 6, is simple if the consumer is part of table TabX, e.g., in a column User. This is in fact how row level security works in commercial database products. Then, the <Condition> is quite generic: "User=current_user" by applying the current_user function in PostgreSQL.

However, it seems to be rather unrealistic that the subscribed consumer already occurs in the column data. It would certainly be more flexible if a subquery could determine the visible records for a consumer. As an important requirement to be taken into account, the approach must avoid a re-compilation of the DDS for any new customer. Moreover, the communication between frontend API-M and DDS or database should at least be reduced because business processes for service subscription have to be implemented, the possibilities of which are limited.

One approach is to implement the functionality in the database by computing the keys of visible records for a table TabX by a table-valued function RowLevelFilter4TabX(user VARCHAR):

```
CASE user WHEN "user1" THEN SELECT t.key FROM TabX t ...
          WHEN "user2" THEN SELECT t.key FROM TabX t ...
```

Hence, row-level filtering can rely on any condition, on any columns or data. Even more, the view TabX_Filter (cf. Figure 6) remains stable by replacing <Condition> with:

```
LEFT OUTER JOIN RowLevelFilter4TabX(current_user) c
          ON c.key = t.key
```

This principle offers complete freedom for a consumer-specific row-level filtering. However, there is a small disadvantage: For each new consumer, the function must be extended in order to add a consumer-specific subquery in the CASE clause. Fortunately, this can be done in the database at runtime without any impact on the DDS and its implementation, and without downtime.

An alternative is to keep the condition in the Privileges table in a textual form and to rely on dynamic SQL to compose an overall query.

6.5   Administration

Some administrative effort is required for the presented approach. To summarize the previous discussion, a common database connect user is required for the DDS first, e.g., in PostgreSQL:

```
CREATE ROLE loginOnly NOINHERIT ...;
```

Furthermore, the Privileges table and the TabX_Filter views must be created. These activities occur only once.

Next, several statements have to be executed for every new consumer <user>:

- CREATE ROLE <user>;
- GRANT <user> TO loginOnly;
- GRANT SELECT ON TabX_Filter TO <user>;

Moreover, a new consumer requires additional records in the Privileges table to specify column access; otherwise default settings apply to them. Finally, the function RowLevelFilter4TabX has to be adjusted in order to implement row-level filtering for a consumer.

In total, the consumer-specific administrative operations are minimal and do not affect the implementation of the DDS.

In principle, the above consumer-specific activities have to be established as part of the subscription workflow of the API-M. DDS can offer a service to execute those tasks in the database. This service can then be used by the workflow process.

In case a specific subscription workflow cannot be defined in the API-M tool, we can let DDS keep a table AllConsumers of consumers who have already accessed the DDS successfully. If a new consumer signs in, s/he does not occur in the AllConsumers table, the consumer will be added and the setup is performed. This principle can also be applied in general to avoid a communication between API-M and database.

6.6   Implementation of the DDS

The DDS is implemented in such a way that various database systems and databases can be handled. The implementation is done with the Spring framework.

A `@Controller` class `DDSApiControler` takes care of accepting requests and delegating every request to some database handlers. This class is the REST controller for the DDS and defines the service endpoint as `@RequestMapping` ("/ `{dbType}/{dbase}/query`") for any service request. The controller analyzes the URI of the DDS by determining the type of database system, i.e., extracting `<DBType>`, `<Database>`, and `<Options>` as well as the SQL query in the request body. Moreover, it performs some basic authentication check to prevent usage beyond API-M. The following is a sketch of the class:

```
@RequestMapping(value = "/{dbType}/{dbase}/query",
                method = RequestMethod.POST)
public @ResponseBody Result executeSelectQuery(
  @PathVariable String dbType, @PathVariable String dbase,
  @RequestParam(value="top",required=false) String top,
  ...
  @RequestHeader HttpHeaders headers)
  throws UnauthorizedException, TechnicalException {
  // set database-specific handler according to <DBType>
  DDSHandlerIF dds = ddsLoader.getDDSHandler(dbType);
  String sql = extract from request body;
  Set<Entry<String, List<String>>> headerList
                             = headers.entrySet();
  // search for x-jwt-assertion:
  for (Entry<String, List<String>> e : headerList) {
    String k = e.getKey();
    if (k.equals("x-jwt-assertion")) {
      String idToken = e.getValue().get(0);
      jwt = JWTParser.parse(idToken);
      claims = jwt.getJWTClaimsSet();
      endUser = claims.getStringClaim  // cf. Figure 5
            ("http://wso2.org/claims/enduser");
      ...
    }
  } // delegate request to handler:
  return dds.executeSelectQuery(sql, metainfo, top,
                               offset, endUser);
}
```

The controller extracts the `endUser` and other data from the security token. While usually returning `HttpStatus.OK`, exceptions such as `UnauthorizedException` (for unauthorized access) and `TechnicalException` (for other technical exceptions) are caught in case of errors and mapped by means of an `@ExceptionHandler` to `HttpStatus.FORBIDDEN` and `HttpStatus.INTERNAL_SERVER_ERROR`, resp.

The internally used method `executeSelectQuery` is defined by an interface `DDSHandlerIF`. Further methods, for example, to invoke stored procedures or to modify data, can be defined in addition. All these methods are implemented in a generic manner by an abstract class `GenericDDSHandler` and database-server specific subclasses. `GenericDDSHandler` concentrates the common functionality and leaves database-specific parts open, while classes such as `Post-greSQLDDSHandler` extends the abstract `GenericDDSHandler` class by handling database-specific parts. Consequently, each specific subclass takes care of specific points such as how to specify pagination in corresponding dialects. PostgreSQL and MySQL use `LIMIT` and `OFFSET` clauses in SELECT statements, while SQL Server uses a `Top n` clause. Hence, a method `String addPagination (String sql, String top, String offset)` has to be overwritten. Further points are to switch the consumer, e.g., by means of SET ROLE in PostgreSQL, the way how to obtain a connection, and to map database-specific error codes to exceptions `UnauthorizedException` and `TechnicalException`. Credentials for connecting to the database are stored in configuration files.

7   Specific Points

7.1   Further Forms of Authentication

The DDS obtains from API Management a new tenant-specific URI for a REST API. Various types of clients, no matter whether an application or a graphical user interface, implemented in any language, can invoke the REST API. However, this API is protected by authentication by means of OAuth tokens. That is, to use the API, a security token is required, and every usage is controlled. As a consequence, the client has to obtain and pass on the security token. The token cannot be hard-coded in the code since it expires after a certain period of time. One option is that the application's user interface asks the user for a valid security token again and again, which is then passed to API-M. Since the token has to be requested by the consumers' web interface of the API-M (cf. Fig. 2), this is a manual interactive action. This is quite cumbersome and pollutes the application with an additional input form.

Another option is to let the application request the token programmatically. In fact, WSO2 offers a corresponding REST API that requires the user and password as provided during sign-up and returns the security token. This is also suboptimal since the application again has to acquire user and password from the consumer in the user interface in order to request the token eventually.

A better solution is to let graphical user interfaces (GUIs) in HTML and JavaScript benefit from an advanced OAuth support. To this end, any GUI that wants to use the DDS can be registered for the DDS by specifying a callback. Whenever the DDS API is invoked from the GUI, WSO2 is implicitly contacted. The callback is used by WSO2 to let a login form pop up in the GUI asking the consumer to authenticate with user and password. Another form asks the user to confirm that the GUI is permitted to act on behalf of the consumer. Thus, this OAuth support provides a better integration of authentication in GUIs.

7.2   SQL Injection

The REST API of the DDS is generic in the sense that arbitrary SQL statements can be passed to the DDS as a string in the request body of a POST request. Any type of SELECT statement can be executed. However, passing arbitrary strings usually lead to SQL injection issues. SQL injection denotes the possibility to place malicious code in SQL statements, e.g., via web page input, and is one of the most common web hacking techniques. SQL injection typically occurs whenever a user form asks for input, e.g., an id or some search topic. If the user fills in some SQL statement or command instead of the expected id or topic, SQL injection occurs and might give users access to parts of the database s/he should not see. Sometimes, the system could even be damaged. For example, quite often a variable will take the user input from a form in the following manner:

```
input = getRequestString("Id");
sqlTxt = "SELECT * FROM Person WHERE Id=" + input;
```

The classical SQL Injection is based on the fact that 1=1 is always true. The original intent of the query and corresponding input form is certainly to search for a person with a given id. But if nothing prevents a user from entering other input than the value for id, the user can enter some dangerous input like "4711 OR 1=1". This obviously leads to a falsified condition WHERE Id=4711 OR 1=1: As the condition is valid and always evaluates to true, all rows from the Person table are returned. Indeed, the damage would be high if the table Person contained passwords. Another common example is to add a UNION clause with an additional subquery of the same structure to another table, as in "4711 UNION SELECT SomeColumns FROM AnotherTable": This extends the original result with some other data from AnotherTable.

How can SQL injection occur in the Data Delivery Service (DDS)? The intention is to allow the execution of arbitrary queries even if "OR 1=1" or a UNION clause is added. There is no need to prohibit those queries because a consumer-specific filtering takes place to restrict the result anyway. And indeed, here lies the – not so obvious – risk potential of SQL injection.

The filtering relies on switching the role for a consumer by means of SET ROLE in the backend internally, and then the role determines what rows and columns become visible. In order to obtain more data than intended and allowed, a user can try to execute "SET ROLE consumer;" in a query to switch the role or user: This would activate the filtering according to another consumer's privileges.

One possibility for a user to achieve this is to issue a composed statement "SET ROLE AnotherRole; SELECT ...;". However, SELECT statements are executed with the executeQuery method of JDBC. Executing the "composed" query, JDBC will return an exception. Hence, SET ROLE does not become effective.

Another intrusion attempt is to execute two commands subsequently in two separate requests: "SET ROLE AnotherRole;" and "SELECT * FROM ...;"

Each of both requests obtains a database connection from a pool. Even if both statements obtain the same connection by chance, the second statement is prefixed with "SET ROLE RealConsumer;" within the DDS backend. There is no real danger.

If the DDS supported DML statements (UPDATE/INSERT/DELETE), we would be supposed to use executeUpdate/execute in JDBC. Again, there are the two options as before. The user can issue a command: "SET ROLE AnotherRole; UPDATE ...;". In this case, the user gets elevated to AnotherRole and is able to execute the UPDATE command. Hence, the DDS has to check the SQL command for any composed queries. If the user issues two commands subsequently, the same (safe) behavior occurs as for SELECT queries.

7.3   Denial of Service

The DDS REST API is flexible as it allows a consumer to execute any SQL statement, more precisely SELECT statement for retrieval – of any complexity. Running very complex queries has an impact on the overall performance, not only for the issuing consumer (who submitted the long-lasting query), but also for other concurrent users. Maleficent consumers might exploit this property for denial of service attacks by issuing a huge amount of complex queries having solely the intention to constrain others. The arising question is how to go against. There are several options:

1. Enforcing throttling is a first option taking benefit from some built-in support by WSO2 (cf. Sect. 7.4). Having throttling configured, throttling will automatically take place if too many requests are submitted by a consumer in a certain period of time. Nonetheless, the negative impact of the very first query cannot be prevented.
2. Indeed, the proposed DDS is just one proposal. Any other form of interface beyond providing full SQL is also possible. For example, an API can offer only pre-defined queries with parameters to be filled out. Thus, complexity and execution times can be controlled within the implementation by defining the predefined queries properly.
3. The internal logic of the DDS is principally able to check the syntax of issued queries and to evaluate their complexity, for instance, in terms of the number of joined tables, aggregated functions, access to huge tables etc. This requires parsing the query, at least rudimentarily. If a query is too complex, e.g., involving too many joins and aggregate functions, the execution can be denied. A corresponding rejection should be clearly stated in the SLAs so that consumers are informed and not aggravated.
4. Indeed, the DDS has full control over any execution. In fact, the response, as shown in Fig. 4, carries the server-side execution time. The elapsed time can be monitored for every user request and put into relation to given SLAs. Consequently, detecting many too long-running executions could lead to a rejection of further requests of a particular consumer. Again, the SLA should state such an advanced throttling.
5. Finally, this is also a matter of billing: The billing model should be defined carefully (see for example [2]) and cover the previously discussed aspects. The number of executions and execution times could have a direct impact on prices. The more

consumption, the higher the bill. Even if consumers are still able to run denial of service attack, they will have to pay for them.

7.4 Throttling

Throttling enables a tenant to restrict the number of requests to an API during a given period of time. Typical situations to apply throttling are:

- to protect an API from denial of service or other types of security attacks (cf. Sect. 7.3);
- to regulate load in order not to overload the underlying infrastructure;
- to offer an API available with different access policies according to SLAs and billing schemes.

WSO2 offers different types of throttling that can be configured by a provider. First, a provider can select one or more subscription tiers from a list during the process of publication. There are a couple of predefined *subscription tiers* with specified limits, here of WSO2 version 2.0.0:

- Bronze: 1000 requests per minute
- Silver: 2000 requests per minute
- Gold: 5000 requests per minute
- Unlimited: unlimited access

Further tiers can be defined by tenant administrators with other limits (see later).

The available subscription tiers are also visible for consumers in the consumers' view. A consumer can see the offered tiers and select one for subscription. According to the selected tier (and their associated limits), the consumer requests are throttled. Usually, the subscription tiers are applied for monetary purposes: The more requests a consumer wants to submit, i.e., the higher limits are required, the more s/he has to pay.

Beyond the subscription tiers, options for advanced throttling policies can be set for both a particular resource and an API to control access by means of advanced filter rules. WSO2 distinguishes between a resource and an API. An API comprises of one or more resources, whereby a resource handles a particular type of request according to the HTTP verb. Accordingly, resource-level throttling limits the usage of a resource such as a specific GET request.

With advanced throttling policies, limits can be specified for requests that satisfy some filter conditions. Filter conditions can use the following properties and their combinations:

- IP address and address range throttles access for a given IP address or address range. This is useful to give internal users (with internal IPs) other quotas than for external ones.
- Filtering requests can also be based on HTTP request headers, e.g., to set a dedicated limit for XML requests as opposed to JSON. That is, a policy can set a dedicated limit for requests with a Content-Type header of application/xml.

- A security token carries a JWT claim (cf. Fig. 5) that contains meta information of a request. All the constituents of a token can be used in filter conditions, e.g., to apply specific limits for a dedicated user or role.
- HTTP GET requests often possess query parameters, for instance, to restrict the search to a specific category. These query parameters can also be used to limit access.

These policies are whitelists. Complex conditions can be specified in the Siddhi query language. Blacklists are also possible to exclude specific IPs from any request. Advanced throttling policies can be applied for API- and resource-level when an API is published.

A tenant administrator can add a new subscription tier such as Platinum or new throttling policies. The limits can be specified either by the number of requests over time (e.g., 15,000 request/min) or the amount of data bandwidth over time (1 MB/min). Furthermore, an action to be taken when a consumer goes beyond the allocated quota can be specified. Custom attribute values can be attached to provide information about the throttling, SLAs, billing etc. These attributes are displayed as key value pairs in the consumers' view. And finally, Billing Plan information characterizes an API as Free, Commercial, or Freemium.

It is important to note that throttling policies cannot only be enforced in the API creation and publishing state, but also at runtime.

In addition to the previous concepts, *burst control* is an orthogonal concept that is useful for controlling the distribution of requests for a given period of time. For instance, a subscription tier might allow for 5,000 requests per hour. Nothing prevents a consumer from submitting all these requests within the first 10 s. In order to avoid such a scenario, burst control can restrict the quota for instance to 5 requests per second at a second level of throttling.

The throttling levels define the quotas for consumers, but they do not protect the backend from getting overloaded. Even if certain limits can be specified in subscription tiers, the number of requests will increase with the number of consumers. However, there is usually a physical capacity that a backend services is able to handle. Even if none of the consumers exceed the allocated quotas, the overall load might raise the capacity of a backend system. To prevent a backend system from getting overloaded, *hard backend throttling* can define an absolute limit for the total number of requests within a given time period that arrive at the backend system. Exceeding this limit leads to a hard stop of service requests.

7.5 Advanced Filtering

As already mentioned, the provider of the DDS is able to offer a service in different tiers. There are predefined tiers such as Gold, Silver, and Bronze, but new tiers like Platinum can be defined, too. With each tier certain SLAs (especially throttling) and information about the price scheme can be specified as part of the offering in order to become visible for potential consumers.

When a consumer subscribes to an API, s/he can select a subscription tier (which has been described by the provider as part of his offering) thus accepting the associated prices and SLAs. The selected tier can also be used to perform user-specific filtering (row/column level, limit on result sets, quality, de-personalization, aggregation, read/write permissions etc.).

A consumer can also subscribe to several tiers. The generation of a security token is then done for each particular tier. During invocation, the tier is part of the security token (cf. Fig. 4). Hence, the implementation of the DDS can access and use it, e.g., by adding TOP(n) to queries by query rewriting in order to limit the result size according to the tier, or do deliver data in different qualities.

Using the tier to control filtering is also possible within the approach. One solution is to concatenate user and tier (both can be extracted from the token) to a single name with some separation symbol in between. This name is then passed to the database instead of the consumer as before. That is, a role for this name has to be created in the database, and both parts of the name have to be extracted from the role in the view.

8   Conclusions

In real life, we are often faced with the situation that developers require real data in order to develop smart algorithms, e.g., applying machine learning techniques. Unfortunately, data owners are not willing to give data away for free. The other way round, data owners might benefit from the newly developed algorithms or their results. But representing their intellectual property, developers want to earn money for their efforts.

This paper tries to solve this dilemma by proposing a marketplace to trade data access and algorithms. Such a marketplace supports both providers of data access or algorithms as well as consumers. While providers are enabled to publish data access as a service at various levels, consumers can browse through the offerings, subscribe to particular services if interested, and use the services after being approved by the provider.

The marketplace provides a platform that enables data providers to generate, advertise, and sell access to APIs (especially for data access), whereas consumers are enabled to browse through the offerings and to purchase access to APIs. There are dedicated provider and consumer web interfaces.

However, it is usually not sufficient to give data access to the whole amount of data. Instead, data access should be granted to portions of data, maybe at varying degree of quality according to the consumer, his/her organization, and agreed service level agreements. Therefore, the paper presents a novel approach to support delivering data in a consumer-specific manner depending on some configuration. That is, some row- and column-level filtering, as suggested in the literature for databases, is integrated into the architecture of the marketplace as an important feature.

Central part of the approach is a Data Delivery Service (DDS). The DDS works as a service to access data by means of the full power of SQL and offers row-/column-level access control for relational database systems in a flexible and configurable manner. Similar approaches for other data sources are also possible, e.g., ontologies like DBpedia (http://wiki.dbpedia.org/) with other APIs like SparQL or general computing services.

The DDS is integrated into the WSO2 API Management tool, which takes care of marketplace functionality. The combination of WSO2 and the DDS eases the implementation of the marketplace. The overall architecture of the approach has been presented in detail in order to discuss the advantages, benefits, and flexibility. In particular, any data access is controlled by an OAuth security concept.

In spite of easing the implementation of a marketplace, WSO2 also infiltrated some major challenges to be solved. In particular, the consumer identity has to be passed from the API Management to the DDS and the database in order to perform filtering. And because of an unknown number and possibly numerous users, a flexible configuration without manual administrative effort becomes mandatory. Those challenges and their solutions are discussed with potential alternatives in detail.

We conducted the development of the marketplace in a funded project in the medical domain [20]. In a concrete project setup, we used the DDS to provide access to a medical i2b2 database (https://www.i2b2.org/). Moreover, we illustrated effectiveness of the overall approach by integrating other value-added services in the way described in this paper in a WSO2-based marketplace. For instance, we developed a value-added service on top of the DDS in order to make data anonymous before returning results. These enhancements to the DDS have also been integrated into the marketplace. In summary, we were able to successfully control access of medical professionals within a clinic, having the potential to extend access, maybe in an anonymous manner, outside a hospital.

In our future work, we first want to evaluate the performance impact of filtering in detail. Moreover, we intend to investigate other types of database interfaces besides SQL. One candidate is OData (http://www.odata.org). OData behaves similar to object/relational mapping tools. Starting with an object model, specified in the Entity Framework or JPA, a high-level REST API is provided the functionality of which nearly reaches the power of SQL. Further plans attempt to check whether the approach is adequate to integrate advanced restrictions given by any type of regulation given by both governments or dictated by other bodies.

Acknowledgements. This research has been supported in part by the KDI project, which is funded by the German Federal Ministry of Economics and Technology under grant number 01MT14001 and by the EU FP7 Diachron project (GA 601043).

References

1. Abramov, J., Anson, O., Dahan, M., et al.: A methodology for integrating access control policies within database development. Comput. Secur. **31**(3), 299–314 (2012)
2. Balazinska, M., Howe, B., Suciu, D.: Data markets in the cloud: an opportunity for the fatabase community. Proc. VLDB Endow. **4**(12), 1482–1485 (2011)
3. Barker, S.: Dynamic meta-level access control in SQL. In: Atluri, V. (ed.) DBSec 2008. LNCS, vol. 5094, pp. 1–16. Springer, Heidelberg (2008). https://doi.org/10.1007/978-3-540-70567-3_1
4. Bertino, E., Jajodia, S., Samarati, P.: A flexible authorization mechanism for relational data management systems. ACM Trans. Inf. Syst. **17**(2), 101–140 (1999)

5. Caires, L., Pérez, J.A., Seco, J.C., Vieira, H.T., Ferrão, L.: Type-based access control in data-centric systems. In: Barthe, G. (ed.) ESOP 2011. LNCS, vol. 6602, pp. 136–155. Springer, Heidelberg (2011). https://doi.org/10.1007/978-3-642-19718-5_8

6. Chaudhuri, S., Dutta, T., Sudarshan, S.: Fine grained authorization through predicated grants. In: 23rd International Conference on Data Engineering (ICDE), pp. 1174–1183. IEEE (2007)

7. Chlipala, A., Impredicative, L.: Static checking of dynamically-varying security policies in database-backed applications. In: The USENIX Conference on Operating Systems Design and Implementation, pp. 105–118 (2010)

8. Corcoran, B., Swamy, N., Hicks, M.: Cross-tier, label-based security enforcement for web applications. In: Proceedings of the 2009 ACM SIGMOD International Conference on Management of Data, pp. 269–282. ACM (2009)

9. Fischer, J., Marino, D., Majumdar, R., Millstein, T.: Fine-grained access control with object-sensitive roles. In: Drossopoulou, S. (ed.) ECOOP 2009. LNCS, vol. 5653, pp. 173–194. Springer, Heidelberg (2009). https://doi.org/10.1007/978-3-642-03013-0_9

10. Fuchs, L., Pernul, G., Sandhu, R.: Roles in information security – a survey and classification of the research area. Comput. Secur. **30**(8), 748–769 (2011)

11. Hohenstein, U., Zillner, S., Biesdorf, A.: Architectural considerations for a data access marketplace. In: Proceedings of the 7th International Conference on Data Science, Technology and Applications (DATA 2018), Porto (Portugal), pp. 323–333 (2018)

12. Jayaraman, K., Tripunitara, M., Ganesh, V., et al.: Mohawk: abstraction-refinement and bound-estimation for verifying access control policies. ACM Trans. Inf. Syst. Secur. (TISSEC) **15**(4), 1–28 (2013). Article No. 18

13. LeFevre, K., Agrawal, R., Ercegovac, V., et al.: Limiting disclosure in hippocratic databases. In: Proceedings of the 13th VLDB, pp. 108–119 (2004)

14. Oracle: Using Oracle Virtual Private Database to Control Data Access. https://docs.oracle.com/cd/B28359_01/network.111/b28531/vpd.htm#DBSEG007

15. Pereira, O., Regateiro, D., Aguiar, R.: Distributed and typed role-based access control mechanisms driven by CRUD expressions. Int. J. Comput. Sci. Theor. Appl. **2**(1), 1–11 (2014)

16. Rizvi, S., Mendelzon, A., Sudarshan, S., Roy, P.: Extending query rewriting techniques for fine-grained access control. In: ACM SIGMOD Conference, pp. 551–562 (2004)

17. Rjaibi, W.: Data security best practices: a practical guide to implementing row and column access control. https://www.ibm.com/developerworks/community/wikis/home?lang=en#!/wiki/Wc9a068d7f6a6_4434_aece_0d297ea80ab1/page/A%20practical%20guide%20to%20implementing%20row%20and%20column%20access%20control

18. Roichman, A., Gudes, E.: Fine-grained access control to web databases. In: Proceedings of the 12th ACM Symposium on Access Control Models and Technologies, pp. 31–40. ACM (2007)

19. Roman, D., Paniagua, J., Tarasova, T., et al.: proDataMarket: a data marketplace for monetizing linked data. In: Demo Paper at 16th International Semantic Web Conference (ISWC 2017), Vienna (2017)

20. Sonntag, D., Tresp, V., Zillner, S., Cavallaro, A., et al.: The clinical data intelligence project. Informatik-Spektrum **39**, 1–11 (2015)

21. Wang, Q., Yu, T., Li, N., et al.: On the correctness criteria of fine-grained access control in relational databases. In: Proceedings of the 33rd International Conference on Very Large Data Bases, pp. 555–566 (2007)

22. Zarnett, J., Tripunitara, M., Lam, P.: Role-based access control (RBAC) in Java via proxy objects using annotations. In: Proceedings of the 15th ACM Symposium on Access Control Models and Technologies, pp. 79–88. ACM (2010)

23. XACML: XACML – eXtensible Access Control Markup Language. http://www.oasisopen.org/committees/tchome.php?wgabbrev=xacml

FPGA vs. SIMD: Comparison for Main Memory-Based Fast Column Scan

Nusrat Jahan Lisa[1], Annett Ungethüm[1], Dirk Habich[1(✉)], Wolfgang Lehner[1], Nguyen Duy Anh Tuan[2], and Akash Kumar[2]

[1] Database Systems Group, Technische Universität Dresden, Dresden, Germany
{nusratjahan.lisa,annett.ungethum,dirk.habich,
wolfgang.lehner}@tu-dresden.de
[2] Processor Design Group, Technische Universität Dresden, Dresden, Germany
{nguyenduyanh.tuan,akash.kumar}@tu-dresden.de

Abstract. The ever-increasing growth of data demands reliable database system with high-throughput and low-latency. Main memory-based column store database systems are state-of-the-art on this perspective, whereby data (values) in relational tables are organized by columns rather than by rows. In such systems, a full column scan is a fundamental key operation and thus, the optimization of the key operation is very crucial. This leads to have compact storage layout based fast column scan techniques through intra-value parallelism. For this reason, we investigated on different well-known fast column scan techniques using SIMD (Single Instruction Multiple Data) vectorization as well as using Field Programmable Gate Arrays (FPGA). Moreover, we present selective results of our exhaustive evaluation. Based on this evaluation, we find out the best column scan technique as per implementation mechanism–FPGA and SIMD. Finally, we conclude this paper via mentioning some lessons learned for our ongoing research activities.

Keywords: Column stores · Scan operation · Vectorization · FPGA · Pipeline

1 Introduction

The big data world challenging continuously on how to manage analytical complex database queries more efficiently along with high-throughput and low-latency. This leads to have fast database architecture. Therefore, to speedup the performance, database systems shifted from disk to main memory. Because storing as well as processing all data in main memory is faster than data stored on disk or on a flash drive [1,4]. Although it is important that all data must fit in main memory. However, main memories are still multi-gigabyte, while disk can be multi-terabyte. Additionally, organizing relational tables in main memory by column rather than by row effects the query performance positively as data stored in uniform pattern [1,20]. Thus, such database systems are called main

© Springer Nature Switzerland AG 2019
C. Quix and J. Bernardino (Eds.): DATA 2018, CCIS 862, pp. 116–140, 2019.
https://doi.org/10.1007/978-3-030-26636-3_6

memory column store system. In order to increase the performance of analytical queries, two key aspects play an important role in this so-called main memory column store database systems. On the one hand, compressed storage layout is easily adaptable to reduce the amount of data [1,9,27]. On the other hand, main memory column stores are also smoothly adaptable for novel hardware features like vectorization using SIMD extensions [17,25], GPUs [8], FPGAs [19,22] or non-volatile main memory [16].

One of the primary operation in such system is *column scan* [7,13,23], because analytical queries usually compute aggregations over full or large parts of columns. Thus, the optimization of the scan primitive is very crucial [7,13,23]. Generally, the task of a column scan is to compare each entry of a given column against a given predicate and to return all matching entries. To efficiently realize this column scan, Lamport et al. [11] came up with the initial idea of intra-data processing on single processor word. Later, Li et al. [13] proposed a novel technique called *BitWeaving* which exploits the intra-instruction parallelism at the bit-level of modern processors. Intra-instruction parallelism means that multiple column entries are processed by a single instruction at once. In these approaches, multiple encoded column values are packed either horizontally or vertically into processor words providing high performance when fetching the entire column value [11,13]. Moreover, on these approaches query processing are happening directly on the packed processor words without unpacking the data items. Therefore, as the authors have shown, the more column values are packed in a processor word, the better scan performance [13].

Unfortunately, the length of processor words is currently fixed to 64-bit in common processors, which limits the performance of the intra-instruction parallelism based column scans. To overcome this limitation and to increase the intra-instruction parallelism, there exists two hardware-oriented opportunities. On the one hand, Single Instruction Multiple Data (SIMD) instruction set extensions such as Intel's SSE (Streaming SIMD Extensions) and AVX (Advanced Vector Extensions) have been available in modern processors for several years. SIMD instructions apply one operation to multiple elements of so-called vector registers at once which reduce the instruction calls. The size of the vector registers ranges from 128 (Intel SSE 4.2) to 512-bit (Intel AVX-512), whereby these registers can be used instead of regular processor words. On the other hand, Field Programmable Gate Arrays (FPGAs) are an interesting alternative which provide a great deal of flexibility. Because it allows to design specialized configurable hardware components with arbitrary processor word sizes.

Our Contribution and Outline: This paper is the extended version of our paper [14], whereby we investigate both hardware-oriented opportunities using different types of column scan mechanism. Based on that, we make the following contributions.

1. In Sect. 2, we briefly recap the fast column scan techniques as foundation for our work.
2. The implementation using SIMD vector registers is discussed in Sect. 3, while Sect. 4 covers our FPGA implementation along with selective results of our

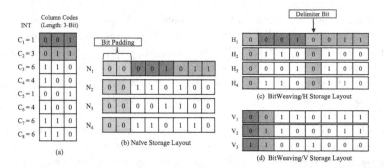

Fig. 1. Storage layout example with (a) 8 integer values with their 3-bit codes, (b) data representation in *Naïve* layout, (c) data representation in *BitWeaving/H* layout and (d) data representation in *BitWeaving/V* layout.

exhaustive evaluation. In particular, we separately evaluate each implementation followed by lessons learned summaries.

3. Finally, we conclude the paper with related work in Sect. 5 and a short conclusion in Sect. 6.

2   Column Scan

Column-oriented database management systems store relational data by columns rather than by rows [4,6]. The advantages are (i) that each column is separately considered and (ii) that the similarity of adjacent column values is preserved. Based on both advantages, the opportunity for compactness and the ability to process multiple column values at once is increased. Thus, the efficient realization of a column scan is an active research topic, whereby each approach consists of two components: (i) storage layout for column values and (ii) scan operation (predicate evaluation) on the proposed storage layout. In this paper, we compare two well-established approaches on two different hardware platforms.

2.1   Naïve

The first column scan technique is the Naïve approach, where each column is encoded with a fixed-length order-preserving code as illustrated in Fig. 1(a). The types of all column values including numeric and string types are encoded as unsigned integer codes [3,9]. The term *column code* refers to the encoded column values. To process multiple column codes in a single processor word during the scan operation, a intra-value parallelism-based compact storage layout is required. Lamport et al. [11] introduced a first approach for this. As shown in Fig. 1(b), column codes are continuously stored horizontally in processor words N_i. As stored codes are fixed in length, the extra unused bits in the processor word are padded with zeros. In our example, we use 8-bit processor words N_1 to N_4, such that two 3-bit column codes fit into one processor word including 2-bit padding per processor word.

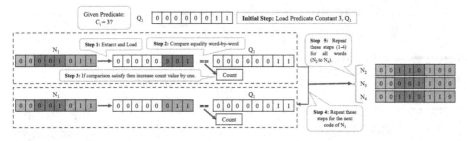

Fig. 2. Equality predicate evaluation using *Naïve/S* technique with extract-load-compare each column code.

During *Predicate Evaluation*, the task of a column scan is to compare each column code with a constant C and to output the number of *Count* indicating how many times the corresponding code satisfies the comparison condition. The predicate evaluation on *Naïve* layout can be done in two ways. *Firstly*, we can evaluate any predicate with simply extract, load and evaluate each (single) code with the comparison condition consecutively, without exploiting code-level parallelism. We named this technique as *Naïve/S*. Figure 2 describes the *equality* check in an exemplary way. The input from Fig. 1(b) is tested against the condition $C_i = 3$. The predicate evaluation steps are as follows:

Initially: Load the predicate constant *3* in word Q_1.
Step 1: Extract one code from N_1 and load in a temporary word.
Step 2: Check equality word-wise between Q_1 and temporary word.
Step 3: If comparison satisfies then increase the value of *Count* by one.
Step 4: Repeat Steps (1 to 3) for the next column code of N_1.
Step 5: Repeat Steps (1 to 4) for the rest of words N_2 to N_4.

Secondly, we can evaluate any predicate directly on the *Naïve* layout with exploiting code-level parallelism. The main advantage of such technique is predicate evaluation is done without decoupling the column codes from a word. We named this technique as *Naïve/M*. Figure 3 illustrated this technique in an exemplary way for the same input and test condition like *Naïve/S*. The detail steps are described as follows,

Initially: Load the *Naïve* layout of predicate constant *3* in Q_1.
Step 1: Check the equality bit-wise of each code between N_1 and Q_1 in parallel. There are 1-bit S_i flag registers for each code of *Naïve* word. For this example (Fig. 3), each word has two (S_1 and S_2) flag registers. If the condition satisfies then set one to S_i flags, otherwise set zero.
Step 2: Perform addition between S_1 and S_2, and store the result in *Count* word.
Step 3: Repeat Step 1 and Step 2 for the rest of words N_2 to N_4 in pipeline manner by overlapping instructions.

In both examples (Figs. 2 and 3), only the second code (C_2) satisfies the predicate, so the resulting *Count* value is one. In order to accelerate column scan,

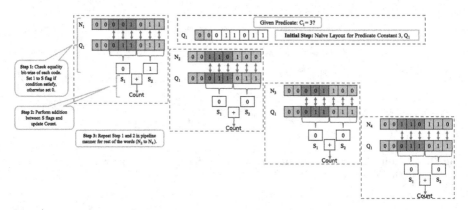

Fig. 3. Equality predicate evaluation using *Naïve/M* technique with directly evaluate on compact words.

Naïve/M technique is good choice than *Naïve/S* for two reason, i) *Naïve/M* technique evaluate predicate directly on the compact word, ii) it is using instruction overlapping mechanism which reduce the number of clock cycles significantly. However, *Naïve/M* technique is difficult to implement on common CPUs using 64-bit processor word, as common 64-bit word do not support intra-instruction parallelism.

2.2   BitWeaving

Our second considered column scan technique is called *BitWeaving* [13]. As illustrated in Fig. 1(a), *BitWeaving* also takes each column separately and encodes the column codes using a fixed-length order-preserving code (lightweight data compression [1,5]), whereby the types of all values including numeric and string types are encoded as an unsigned integer code [13]. To accelerate column scans, BitWeaving technique introduced two types of storage layouts along with an arithmetic framework instead of comparisons for predicate evaluations: *BitWeaving/H* and *BitWeaving/V* [13].

BitWeaving/H. In the storage layout of *BitWeaving/H*, the column codes of each column are viewed at the bit-level and the bits are aligned in memory in a way that enables the exploitation of the intra-cycle (intra-instruction) parallelism for the predicate evaluation. As illustrated in Fig. 1(c), column codes are continuously stored in processor words H_i, where the most significant bit of every code is used as a delimiter bit between adjacent column codes. In our example, we use 8-bit processor words H_1 to H_4, such that two 3-bit column codes fit into one processor word including one delimiter bit per code. The delimiter bit is used later to store the result of a predicate evaluation query.

To efficiently perform column scans using the *BitWeaving/H* storage layout, Li et al. [13] proposed an arithmetic framework to directly execute predicate

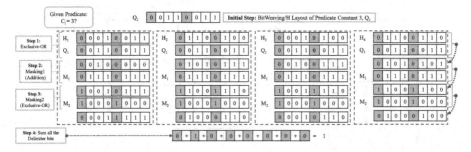

Fig. 4. Equality predicate evaluation using *BitWeaving/H* technique [13].

evaluations on the compressed data. There are two main advantages: (i) predicate evaluation is done without decompression and (ii) multiple column codes are simultaneously processed within a single processor word using full-word instructions (intra-instruction parallelism) [13]. The supported predicate evaluations include equality, inequality, and range checks, whereby for each evaluation a function consisting of arithmetical and logical operations is defined [13]. Figure 4 highlights the *equality* check in an exemplary way. The input from Fig. 1(c) is tested against the condition $C_i = 3$. Then, the predicate evaluation steps are as follows:

Initially: Load the *BitWeaving/H* layout of predicate constant *3* in Q_1.

Step 1: Exclusive-OR operations between the words (H_1, H_2, H_3, H_4) and Q_1 are performed.

Step 2: Masking1 operation (*Addition*) between the intermediate results of Step 1 and the M_1 mask register (where each bit of M_1 is set to one, except the delimiter bits) is performed.

Step 3: Masking2 operation (Exclusive-OR) between the intermediate results of Step 2 and the M_2 mask register (where only delimiter bits of M_2 is set to one and rest of all bits are set to zero) is performed.

Step 4: Add delimiter bits to achieve the total count (final result).

The output is a result bit vector, with one bit per input code that indicates if the code matches the predicate on the column. In the example of Fig. 4, only the second code (C_2) satisfies the predicate which is visible in the resulting bit vector.

BitWeaving/V. In *BitWeaving/V*, the codes are stored vertically across several processor words [13], such that one word contains one bit of several codes. Figure 1(d) shows how the column codes from Fig. 1(a) are stored in the *BitWeaving/V* layout. The words V_i are 8-bit long. The bits of the first number C_1 are stored at the first position of each word, the bits of the second number are stored at the second position, and so on. This way, eight 3-bit codes can be stored across three 8-bit words.

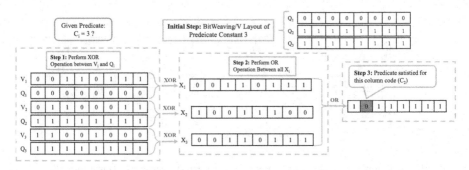

Fig. 5. Equality predicate evaluation using *BitWeaving/V* technique [13].

To evaluate predicate in this layout, we consider the restriction operation *Equality*. However, any kind of (restriction type) predicate evaluation can be perform in this layout. Figure 5 illustrated the *Equality* check predicate evaluation for *BitWeaving/V* in a exemplary way. In the example, we evaluate the column codes C_i for an equality with 3. The necessary steps are:

Initially: Predicate constant *3* is loaded as *BitWeaving/V* layout $(Q1, Q2, Q3)$.

Step 1: XOR operations are performed between *BitWeaving/V* layout based words and predicate constant as follows,

$$X_1 = V_1 \oplus Q_1$$
$$X_2 = V_2 \oplus Q_2$$
$$X_3 = V_3 \oplus Q_3$$

Step 2: Performed bitwise *OR* operations between (X_1, X_2, X_3).

Step 3: In the result word, there is only one position set to *0*. That means, the example condition is satisfied for only one column code and the total count value is one.

2.3 Summary

With the increasing demand for in-memory data processing, there is a critical need for fast scan operations [7,13,23]. The *Naïve/M* and *BitWeaving* techniques addresses this need by packing multiple codes into processor words and applying full-word instructions for predicate evaluations. As shown in [13], *BitWeaving* techniques achieved significant improvement over *Naïve/S* due to its intra-instruction parallelism mechanism. However, they do not consider *Naïve/M* technique. That defines explicitly the more codes are packed in processor words the better performance can be achieved. Unfortunately, processors words in all common CPUs are currently fixed to 64-bit in length. To further speedup *BitWeaving* or *Naïve/M*, larger processor words would be beneficial. To realize larger processor words, we have two hardware-oriented alternatives: (i) vector registers of SIMD extensions or (ii) Field Programmable Gate Arrays (FPGAs). We consider *BitWeaving* and *Naïve/M* approaches for both hardware-oriented alternatives. Both alternatives are discussed in the following sections in detail.

3   SIMD-Implementation

One hardware-based opportunity to optimize column scans is provided by vectorization using SIMD extensions (Single Instruction Multiple Data) of common CPUs. We developed SIMD-implementations for *Bitweaving/H*, *Bitweaving/V*, and *Naïve/M*. *Naïve/S* is intentionally left out because this equals a SIMD-Scan [23] when it is extended to SIMD. A comparison between a SIMD-Scan and the original *BitWeaving* variants has already been done by Li and Patel [13]. In the remainder of this chapter, we shortly introduce the system we used and its SIMD extensions. Then we explain our implementations in detail. Finally, all approaches are compared in an evaluation.

3.1   Target System

A SIMD implementation requires a system with the corresponding registers and instructions. Generally, SIMD instructions apply one operation to multiple elements of so-called vector registers at once. For a long time, the vector registers were 128-bit in size. However, hardware vendors have introduced new SIMD instruction set extensions operating on wider vector registers in recent years. For instance, Intel's Advanced Vector Extensions 2 (AVX2) operates on 256-bit vector registers and Intel's AVX-512 uses even 512-bit for vector registers. The wider the vector registers, the more data elements can be stored and processed in a single vector. Additionally to an increased register size, each new vector extension comes with new instructions, e.g. gather-instructions were first introduced in AVX2. AVX-512 consists of several instruction sets, each providing different functionality, e.g. conflict detection or prefetching.

For the evaluation of our SIMD-implementation, we used an Intel Xeon Gold 6130 with DDR4-2666 memory offering SIMD extensions with vector registers of sizes 128-, 256-, and 512-bit (SSE, AVX2, and AVX-512). This system offers the AVX-512 Vector Length Extensions (VL), which provide most AVX-512 intrinsics for 128-bit and 256-bit registers, that would otherwise only work with 512-bit registers. There is a 32 KB L1 cache for instructions and 32 KB L1 for data. The L2 cache is 1 MB and the LLC (Last Level Cache) is 22 MB. The CPU runs at a base frequency of 2.1 GHz. It has 4 sockets, each containing 16 cores with up to two hyperthreads per core. However, the idea is to observe the influence of the different vector layouts and sizes, not the influence of multiple memory channels or CPU cores. Thus, all benchmarks are single threaded.

3.2   Implementation Details

The SIMD-implementation shows different challenges depending on the evaluation algorithm to be applied. For instance, *Naïve/M* could be implemented easily in regular registers. However, this would not be efficient because single bits cannot be addressed. This introduces an overhead to test whether a set of arbitrary bits, which may or may not be aligned within byte boundaries, is set

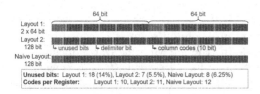

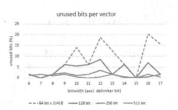

Fig. 6. Different variants to arrange column codes in a vector register.

Fig. 7. Percentage of unused bits per vector register depending on the vector layout [14].

or not. A SIMD implementation has to solve this with a limited number of available instructions. Furthermore, while *BitWeaving/V* is trivially extended from the original approach to vector sizes, *BitWeaving/H* either has to make compromises in the usage of the registers, or work around the fact that the instruction set does not offer a full adder for numbers larger than 64-bit.

Naïve/M

Vector Storage Layout: The *Naïve* storage layout can easily be adapted to vector registers. Figure 6 shows different layouts in an exemplary way for 128-bit registers and 10-bit codes. The layouts for *BitWeaving/H* will be discussed in Sect. 3.2. The *Naïve* layout stores all codes consecutively in a register. Since codes are not spread across two registers, some bits of a register can remain unused. However, the *Naïve* layout is more compact than the *BitWeaving/H* layouts, because there are no delimiter bits.

Predicate Evaluation: The most simple predicate evaluation, which can directly be performed on data in the *Naïve* layout, is an equality check. For such an evaluation, two tasks have to be solved: (1) a bit-wise equality check between the input data and the predicate, and (2) a check for all code words in the input, if all bits of the comparison from step 1 are set. While task one can simply be done by applying an exclusive OR and negating the outcome, task two requires an additional bit-mask to filter the bits of each code word and perform the comparison. This is because we cannot explicitly access arbitrary bits of a vector register. The exact procedure for 128 bit is as follows:

1. Load the predicate in *Naïve* layout with `_mm_loadu_si128`
2. Load data in naïve layout (*input*) with `_mm_loadu_si128`
3. Perform bitwise XOR on the registers loaded in step 1 and step 2 with `_mm_xor_si128`
4. Negate the result from step 3. Perform a bit-wise AND with a vector, where the bits at the position of the current code are set to 1 and all other bits set to 0 (*filter*). `_mm_andnot_si128` performs both operations.
5. Compare the result from step 4 with *filter* using `_mm_cmpeq_epi32_mask`. The result is an 8-bit mask with the first four bits set to one if both vectors are equal.

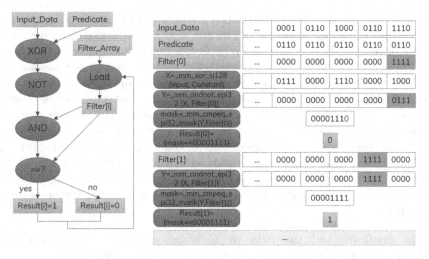

Fig. 8. Evaluation of data stored in the *Naïve* layout. The data is not extracted like in a SIMD-Scan but evaluated directly. The filter array is a precomputed array of vectors, where each vector acts as a filter to zero out every element except for one code word. On the left side, a graph shows the steps necessary for evaluating one input register. The right side shows this in an exemplary way for the first two code words of an input register. In this example, the bit-width of the codes is 4 and the vector width is 128 bit.

6. Compare the result from step 5 with an 8-bit number where the first four bits are set to one.
7. If all codes in *input* have been processed, repeat from step 2, else repeat from step 4 with the next code in *input*.

This procedure can be ported to 256 and 512 bit by simply renaming the intrinsics accordingly. For a better understanding, Fig. 8 illustrates the whole process for a single input register and provides an example for evaluating the first two codes in this register using SSE.

BitWeaving/H

Vector Storage Layouts: A straightforward way to implement *BitWeaving/H* using vector extensions is to load several 64-bit values containing the column codes and delimiter bits into a vector register. In this case, the original processor word approach is retained as proposed in *BitWeaving*. This vector layout is shown as *Layout 1* in Fig. 6. However, this method does not use the register size optimally. For instance, in a 128-bit register, there is space for 11 column codes with a bit width of 10 and their delimiter bits (see Fig. 6 *Layout 2*), but *Layout 1* can only hold 10 codes. In *Layout 2*, we treat the vector register as a full processor word and arrange the column codes according to the vector register size. Figure 7 shows the percentage of unused register space for different register sizes and both layouts, where the dashed line shows the usage for *Layout 1* and the remaining lines for *Layout 2*. As we can see, *Layout 2* makes better use of the vector register. For our evaluation in Sect. 3.3, we implemented both layouts.

Predicate Evaluation: Like in the original approach, the query evaluation on data in the *BitWeaving/H* layout in vector registers consists of a number of bit-wise operations and one addition. The exact bit-wise operations and their sequence depends on the comparison operator. For instance, a *smaller than* comparison or an *equality* check requires XOR operations and an addition as shown in Sect. 2.2. For counting the number of results quickly, an AND is also necessary. For 512-bit registers, this is realized by using AVX-512 intrinsics. The following steps are necessary for a *smaller than* comparison if the data is using the vector *Layout 1* (see Fig. 6):

1. The predicate and the data in *BitWeaving/H* layout is loaded with `_mm512_loadu_si512`. The constraint must only be loaded once.
2. The bit-wise XOR is performed with `_mm512_xor_si512`.
3. The addition is performed with `_mm512_add_epi64`.
4. Optional: To set only the delimiter bits, an AND between the precomputed inverted bit-mask and the result from step 3 is performed with `_mm512_and_si512`.
5. Optional: For counting the number of set delimiter bits `_mm512_popcnt_epi64` is applied.
6. Optional: The result from step 5 can be further reduced by adding the individual counts with `_mm512_reduce_add_epi64`.
7. Finally, the result is stored with `_mm512_storeu_si512`. If only the number of results is required, this step can be skipped. Afterwards, a new iteration starts at step 1.

Note that the SIMD intrinsics for step 5 and 6 do not exist for 128-bit and 256-bit registers. In these cases, the result is written back to memory and treated conventionally, i.e. like an array of 64-bit values.

These steps work for *Layout 1* but not for *Layout 2*. This is because in step 3, a full adder is required. However, this functionality is supported for words containing 16, 32, or 64 bits, but not for 128, 256, or 512 bits. Hence, this adder must be implemented by the software.

Full Adder for Large Numbers: While *Layout 2* uses the size of the vector register more efficiently, it comes with a major drawback: There is no full adder for more than 64 bit on recent CPUs. The evaluation with *BitWeaving/H* uses mainly bit-wise operations but one addition is necessary in all evaluations, i.e. equality, greater than, and smaller than. To realize this addition for 128-, 256-, or 512-bit, there are two different ways: (a) the addition is done by iterating through the bits of the summands and determining and adding the carry bit in every step, and (b) only the carry at the 64-bit boundaries is determined and added to the subsequent 64-bit value. Option (a) requires sequential processing and cannot

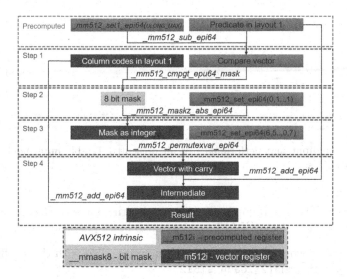

Fig. 9. A software adder for large numbers using AVX-512 intrinsics. For *BitWeaving*, the two summands are the predicate and the column codes. This approach can easily be adapted for 128 and 256 bits [14].

be implemented in a vectorized way. Thus, we chose option (b). The exact steps for option 2 are shown in Fig. 9 for 512-bit vector registers:

1. Since the result of the addition of two 64-bit values is also 64-bit, a potential overflow cannot be determined directly. Instead, we subtract one summand from the largest representable number and check whether the result is larger than the other summand. If it is smaller, there is a carry. This can be done vectorized. The output of the comparison between two vector registers containing unsigned 64-bit integers is a bit-mask.
2. The bit-mask resulting from step 1 is used on a vector containing only the decimal number 1 as 64-bit value at every position.
3. A carry is always added to the subsequent 64-bit value. For this reason, the result from step 2 is shifted to the left by 64-bit. This is realized by intrinsics providing a permutation of 64-bit values.
4. Finally, the two summands and the result from step 3 are added.

All steps can be done using AVX-512 intrinsics. If one of the summands is a constant, like the predicate in *BitWeaving*, the subtraction in step 1 can be precomputed.

BitWeaving/V. The implementation of vectorized *BitWeaving/V* is straightforward because all needed functionality is provided by the SIMD intrinsics of SSE, AVX2, and AVX-512. The layout stays exactly the same as described in Sect. 2.2. In our case, the processor words V_i are 128-bit, 256-bit, or 512-bit long. The number of necessary words for a segment equals the number of bits

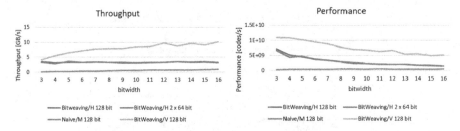

Fig. 10. Throughput and performance of all presented 128-bit implementations for growing code sizes.

per code word. The evaluation is also done as described in Sect. 3.2. For instance, an equality check using AVX-512 requires the following steps for each segment:

1. Load the first word of the segment and a vector filled with the 1st bit of the predicate with `_mm512_loadu_si512`.
2. Perform a bit-wise XOR on the registers loaded in step 1 with `_mm512_xor_si512`
3. Invert result from step 1. Perform a bit-wise AND with a 1-vector if it is the first word of the segment, perform bit-wise AND with the result from the last iteration otherwise. The inverting and bit-wise AND are done with `_mm512_andnot_si512`.
4. Repeat from step one with next word of the segment and the next bit of the predicate.

3.3   Evaluation and Summary

In the evaluation, we want to observe the influence of the different vector layouts and sizes, not the influence of multiple memory channels or CPU cores. Thus, all benchmarks are single threaded. All measurement values are averaged over ten runs.

Overview. Figure 10 shows a comparison of the performance (codes/s) and the throughput (GB/s) of all implementations using 128 bit. As expected, the *Naïve* layout provides the lowest throughput and performance. The time needed for evaluating every code in a register individually cannot make up for the slightly better usage of the available bits. The two layouts of *BitWeaving/H* do not show any significant differences but perform better than the *Naïve/M* approach.

Finally, *BitWeaving/V* shows the highest throughput and performance as could be expected since it is the approach with the least operations, which have to be performed while the input vector layout is more compact than in *BitWeaving/H*. Moreover, it has the smallest output size, resulting in less store operations, i.e. the bits containing the evaluation result are stored consecutively in the result register. *BitWeaving/V* is the only implementation, where the throughput increases when the code size increases, while *Naïve/M* and *BitWeaving/H* show

	Naïve/M	BitWeaving/H								BitWeaving/V		
Register size [bit]	128	64	128	2 x 64	256	4 x 64	512	8 x 64	128	256	512
Throughput-wise	☒	○	○	○	○	○	○	○	☑	☑	☑
Performance-wise	☒	○	○	○	○	○	○	○	☑	☑	☑

Fig. 11. Comparison of the implemented column scan techniques.

a constant throughput. A reason for this behaviour is that in *BitWeaving/V*, the number of result bits per input bit decreases when the code size increases because more data is needed to compute a result. This leads to less store operations for the same amount of input data. For instance, if the bit width of the codes is 3, one segment consists of 3 processor words. Thus, the result, i.e. one processor word, is written back after these 3 processor words have been evaluated. But if the bit width is 15, there are 15 processor words, which are evaluated before one processor word is written back to memory. At the same time, the performance decreases for all *BitWeaving* approaches while the size of the codes increases. In *BitWeaving/H*, this is because less codes fit into one processor word when the code size increases. In *BitWeaving/V*, it takes more operations before a result is computed as explained before. Before going into detail, Fig. 11 shows an overview of all implementations. While *BitWeaving/V* stays clearly on top of the other approaches, it also shows some variation between the different vector sizes. *BitWeaving/H* is less influenced by the vector size.

BitWeaving/H. For codes containing 3 bits and a delimiter bit, the non-optimized 64 bit implementation achieves a throughput of 2.9 GB/s, which equals a performance of almost 58e+8 codes per second.

The results for 3-bit column codes for all different horizontal vector layouts are shown in Table 1. All values are averaged over 10 runs. The results show, that there is a performance gain when using the vectorized approach, but it is not as significant as expected. For instance, we would expect a 100% speed-up when changing from 64 to 128 bits since we can process twice the data at once. Unfortunately, the throughput and the performance increase only by 14%. Moreover, it even decreases when changing from 256 to 512 bits for both vector layouts. However, these numbers can only provide a rough estimation since the throughput varies by up to 0.5 GB/s between the individual runs.

Figure 12 shows the throughput and performance for all implemented *BitWeaving/H* versions and different code bitwidths. For comparison, we also implemented a scalar 64-bit *BitWeaving/H* version without any further optimization for special cases, such that the predicate evaluation is always executed in the same way.

The differences between the vectorized implementations and the scalar implementation becomes even smaller when the code size increases while the throughput oscillates between 2.5 GB/s and 4 GB/s for all versions (see Fig. 12). There is a mere tendency of the 256-bit implementations to provide the best performance

Table 1. Evaluation results on Intel Xeon Gold 6130, 3 bits per code, average over 10 runs [14].

Vector layout	Throughput (GB/s)	Performance (Codes/s)
None (64-bit) (baseline)	2.9	57.8e+8
2X64-bit (Layout 1)	3.3	65.7e+8
4X64-bit (Layout 1)	3.5	69.3e+8
8X64-bit (Layout 1)	2.9	57.6e+8
128-bit (Layout 2)	3.6	71.6e+8
256-bit (Layout 2)	3.6	72.4e+8
512-bit (Layout 2)	2.9	58.9e+8

in average and for the 512-bit versions to provide the least performance. Nevertheless, the insignificance of the differences cannot be explained with the query evaluation itself. To find the bottleneck, we deleted the evaluation completely, such that only the vectorized load and store instructions were left. Then, we measured the throughput again and received results between 3 GB/s and 4 GB/s. A simple `memcopy` had a stable performance around 4.5 GB/s. Hence, in contrast to the naive implementation, the vectorized implementations are bound by the performance of loading and storing data, while the peak throughput cannot become larger than 4.5 GB/s.

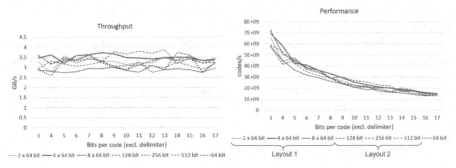

Fig. 12. Throughput and performance for *BitWeaving/H* [14].

BitWeaving/V. The performance and throughput of all implemented *BitWeaving/V* versions for different code sizes are shown in Fig. 13. Contrary to *BitWeaving/H*, there is a clear increase of performance and throughput when the register size increases. A reason for this is the already mentioned smaller output size. Unlike in *BitWeaving/V*, in the horizontal approach, there is a padding between the result bits, which is as wide as a code word. To get these result bits,

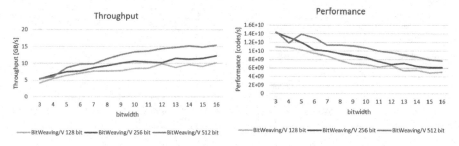

Fig. 13. Throughput and performance for *BitWeaving/V*.

the whole vector word has to be extracted to several regular registers, where the bits can be shifted together, or even written back to memory completely if there are not enough registers. Since it is common for CPU cores to have only 16 general purpose registers, this worst-case is the usual case. However, *BitWeaving/V* does not have such a padding, which makes the output more compact and reduces store operations. This relaxes the memory bandwidth bottleneck to a certain degree. This is especially obvious in the throughput for larger code sizes, where there are more input registers processed before the output register is written back. The performance decrease for 512 bit at a code size of 4 bit is reproducible. It comes with a throughput, which is not increased as much as expected. We did not find an explanation for this in the algorithm itself, especially because it only occurs for 512 bit. A possible reason is a fail of the optimizer during compilation. To test this theory, we compiled the exactly same source code with icc, whereas we were using gcc before. The results did not show the decrease at 4 bit. Instead, there is a peak at 10 bit and the overall increase is less steady. Thus, it is safe to assume that these outliers are caused by the compiler rather than the implementation or the hardware.

4 FPGA-Implementation

Besides the implementation by means of wider vector registers, the second hardware based implementation possibility is the use of Field Programmable Gate Arrays (FPGAs). FPGAs are integrated circuits, which are re-configurable after being manufactured. More specifically, a hardware description language, e.g., Verilog, is used to describe the hardware modules. This description is then translated via several steps to an implementation for the FPGAs. From the perspective of intra-value or intra-instruction parallelism based storage layout, the advantage of FPGAs is that we are able to use an arbitrary length of processor word in the custom hardware design.

4.1 Target System

Modern FPGAs are integrated with MPSoC (multiprocessor system on chip) architectures. The Xilinx® Zynq UltraScale+™ platform—our target FPGA

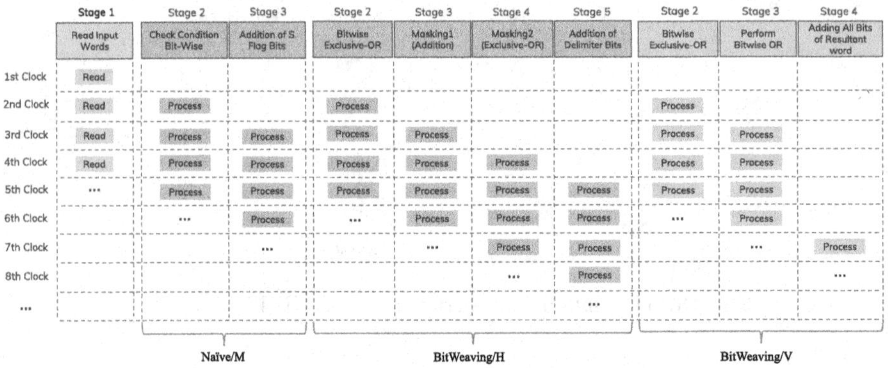

Clock	Stage 1 — Read Input Words	Stage 2 — Check Condition Bit-Wise	Stage 3 — Addition of S Flag Bits	Stage 2 — Bitwise Exclusive-OR	Stage 3 — Masking1 (Addition)	Stage 4 — Masking2 (Exclusive-OR)	Stage 5 — Addition of Delimiter Bits	Stage 2 — Bitwise Exclusive-OR	Stage 3 — Perform Bitwise OR	Stage 4 — Adding All Bits of Resultant word
1st Clock	Read									
2nd Clock	Read	Process		Process				Process		
3rd Clock	Read	Process	Process	Process	Process			Process	Process	
4th Clock	Read	Process	Process	Process	Process	Process		Process	Process	
5th Clock	...	Process	Process	Process	Process	Process	Process	Process	Process	
6th Clock		...	Process	...	Process	Process	Process	...	Process	
7th Clock			...		...	Process	Process		...	Process
8th Clock						...	Process			...
...							...			

| Naïve/M | BitWeaving/H | BitWeaving/V |

Fig. 14. Pipeline-based PE for different intra-instruction parallelism based column scan techniques.

system—is such a system containing not only Programmable Logic (PL) based FPGA, but also has MPSoC-based Processing System (PS) in particular having four ARM® Cortex-A53 cores with 32 KB of L1 instruction cache resp. 32 KB data cache per core and a 1MB shared L2 cache. The main memory consists of two memory modules (DDR4-2133) with the accumulated capacity of 4.5 GB. Although the main memory of our targeted FPGA platform has limitations regarding capacity and bandwidth compared to modern Intel systems, the flexibility to prepare any type of custom hardware and the high parallelism criteria of FPGAs are very beneficial to overcome these challenges.

4.2 Implementation Detail

Inside the PL area of FPGAs, we can develop Processing Element (PE) modules for any type of predicates using Configurable Logic Block (CLB) slices, where each CLB slice consists of Look-up Tables (LUTs), Flip-Flops (FFs), and cascading adders [22]. As illustrated in Fig. 14, the stages of PEs are processing words in pipeline manner through overlapping instructions, whereby we developed 3-stage, 5-stage and 4-stage pipeline-based PE for equality check predicate evaluation on the basis of *Naïve/M*, *BitWeaving/H* and *BitWeaving/V* techniques as introduced in Figs. 3, 4 and 5, respectively. All PEs have a common *Stage 1* of reading data words from main memory (see Fig. 14). Rest in every stages a specific task is performed as shown in Fig. 14, whereby the stages for different techniques are grouped by colors. The detail of *Naïve/M* pipeline stages are:

Stage 2: Check equality condition bit-wise and set S flag values according to the condition satisfying result,

Stage 3: Perform addition between S flags in order to count the matched column codes.

Then, the detail of *BitWeaving/H* pipeline stages are:

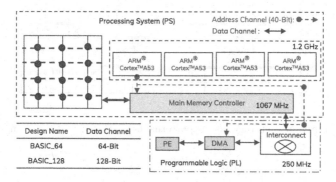

Fig. 15. Basic architecture [14].

Stage 2: Executing bit-wise Exclusive-OR operations,

Stage 3: Masking operations (Addition),

Stage 4: Masking operations (Exclusive-OR) using predefined mask registers to prepare the output word,

Stage 5: Adding delimiter bits of output words in order to count the matched column codes.

Finally, the detail of *BitWeaving/V* based pipelines are:

Stage 2: Executing bit-wise Exclusive-OR operations,

Stage 3: Executing bit-wise OR operations,

Stage 4: Adding all bits of previous stage resultant words in order to count the matched codes (this stage would execute after every w cycles, whereas w is the width of column code).

For all cases, we write only the final output word of count to the main memory. This is not shown in Fig. 14 as it is a non-pipeline stage which executes once only. Therefore, the total number of cycles for *Naïve/M*, *BitWeaving/H* and *BitWeaving/V* is $(n + 3)$, $(n + 5)$ and $(n + 4)$, respectively, where n is the total number of input words.

Basic Architecture. We started with developing 64-bit word based hardware design as Basic architecture (BASIC_64) and subsequently increased the word width to 128-bit (BASIC_128) (see Fig. 15). In this architecture, we use Direct Memory Access (DMA) between the main memory and the PE, in order to reduce the load of the ARM core and to reduce the latency of accessing the main memory. We prepared basic architecture based designs having either *Naïve/M* or *BitWeaving/H* or *BitWeaving/V* technique based PE, whereas each design is processing either 64-bit or 128-bit words.

Hybrid Architecture. The main challenge comes up when the word to be processed become larger than 128-bit, because the width of the data channel

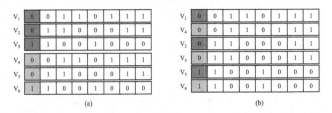

Fig. 16. *BitWeaving/V* storage layout patterns, (a) for basic and (b) for hybrid architectures.

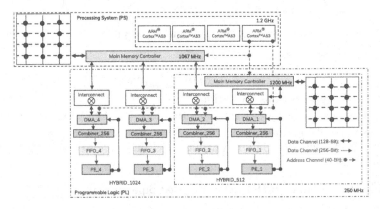

Fig. 17. Hybrid architecture [14].

of the main memory can only be extended up to 128-bit although the PEs are capable to handle word sizes beyond 128-bit. To tackle this challenge, we developed a hybrid architecture based on multiple DMAs, where each DMA is accessing the main memory via an independent data channel. As a consequence, we replicate our PE and DMA a few times depending on the number of available main memory data channels.

Moreover, two main memory modules are available on our targeted FPGA platform as mentioned earlier: one is connected with the PS and the other one is connected to the PL. The PS part main memory has four data channels, while the PL part has only one. However, maximum channel width is 128-bit. So, maximum five times of 128-bit words can be processed in parallel by using multiple main memory modules. However, having maximum number of data channels in a design saturates the bandwidth of main memory. Therefore, we can prepare another custom hardware module, whereas 128-bit words can be combined into larger words. Thus, we implemented and replicated a custom combiner (namely Combiner_256) to combine two 128-bit words to produce 256-bit word. This introduces another stage in each proposed pipeline design, such that each PE is processing a 256-bit word in each clock cycle. Such a combiner can easily adoptable in *Naïve/M* and *BitWeaving/H* techniques based hardware designs as they stored codes in words horizontally rather than vertically like *BitWeav-*

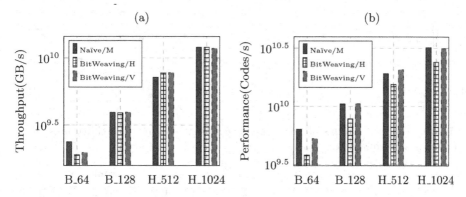

Fig. 18. Analysis on (a) Throughput-wise, (b) Performance-wise for basic and hybrid architectures using different column scan techniques (3-Bit Per Code).

ing/V. Therefore, for *BitWeaving/V*, the input words are stored alternatively rather than sequentially as illustrated in Fig. 16(b) for 3-bit column codes, so that combiner can merge two words perfectly without breaking the sequence of codes. However, we keep the as usual storage pattern of *BitWeaving/V* for basic architecture as described in Sect. 2.2 (see Fig. 16(a)).

In addition, we use appropriate depth based FIFO between the combiners and the PEs to synchronize IO transmission between PEs and main memory, whereas main memory is using stream-based data transmission. This avoids an overflow of the DMA buffer. By mixing all the above mentioned concepts, we prepared hybrid architecture based designs as HYBRID_512 and HYBRID_1024, to process two and four times of 256-bit word in parallel in order to make 512-bit and 1024-bit words, respectively for all techniques (see Fig. 17).

4.3   Evaluation and Summary

Experiments are evaluated using two main metrics: throughput (GB/s) and performance (Codes/s). Although in our previous work, we evaluated energy consumption metric as estimated energy and actual energy for codes per joule on *BitWeaving/H* scan [14]. But in these evaluations, we did not consider energy consumption due to having same behaviour like performance as it depends on codes. In addition, we showed that, our proposed basic and hybrid architectures win over ARM-based evaluations for *BitWeaving/H* scan [14]. Therefore, these evaluations are targeted to analysis the behaviour between *Naïve/M*, *BitWeaving/H*, *BitWeaving/V* column scan techniques for basic and hybrid architectures. We evaluated with BASIC_64, BASIC_128, HYBRID_512 and HYBRID_1024 designs for *Naïve/M*, *BitWeaving/H* and *BitWeaving/V* scan techniques, whereby Fig. 18 shows the results for 3-bit column codes (excluding delimiter bit for *BitWeaving/H* scan) with equality check predicate.

We started with BASIC_64 design based evaluations and found that, *Naïve/M* provides higher throughput than *BitWeaving* techniques (see Fig. 18(a)).

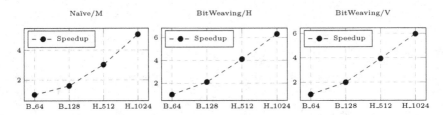

Fig. 19. Analysis in terms of Speedup between basic and hybrid architectures for all column scan techniques.

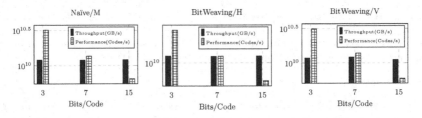

Fig. 20. Analysis on HYBRID_1024 design using different column scan techniques for different number of bits per code.

Because it able to execute on 300 MHz frequency due to having simple logic instruction based technique, whereas others execute on 250 MHz. However, this scenario changed for BASIC_128, HYBRID_512 and HYBRID_1024 based designs, where we achieved approximately same throughput for all techniques (see Fig. 18(a)) as the frequency of these designs are identical. Moreover, different number of total clock cycles of PEs for different techniques as shown in Sect. 4.2 do not effect the throughput due to its pipeline mechanism. In the hybrid architectures-based designs data words are uniformly distributed among the PEs. In addition, the hybrid architecture based designs are processing beyond 256-bit width based data words through multiple main memory data channels and also flexible to use additional hardware (i.e., Combiner_256, FIFO), which is not available on BASIC_64 and BASIC_128 designs. As a consequence, for all techniques, HYBRID_1024 gives the peak throughput of approx. 12 GB/s, whereas three data channels from PS part main memory and one data channel from PL part main memory are used. Although the PS part main memory have maximum four data channels. But using the maximum number of channels in parallel saturates the bandwidth of PS part main memory. So, in HYBRID_1024 we used multiple main memories in order to have four individual data channels.

Performance-wise evaluation varies between different techniques. *BitWeaving/H* provides always less performance in terms of codes per second among all techniques (see Fig. 18(b)). In BASIC_64 design, *Naïve/M* provides the highest performance (see Fig. 18(b)). However, rest in all designs the performance become marginal between *Naïve/M* and *BitWeaving/V* (see Fig. 18(b)). There are two reasons. On the one side, the number of *bit padding* increases in *Naïve/M*

Table 2. Resource utilization for HYBRID_1024 designs.

Scan tech	LUTs (%)	FFs (%)
Naïve/M	12.89	8.64
BitWeaving/H	13.68	9.5
BitWeaving/V	13.99	9.15

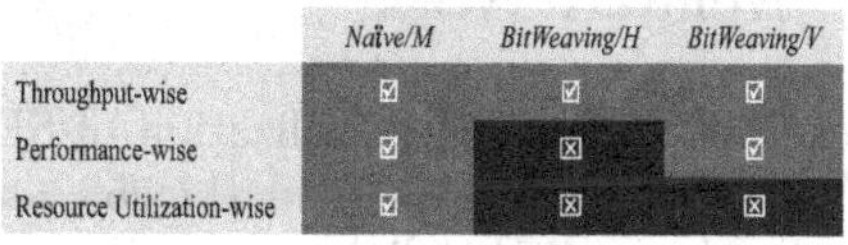

	Naïve/M	BitWeaving/H	BitWeaving/V
Throughput-wise	☑	☑	☑
Performance-wise	☑	☒	☑
Resource Utilization-wise	☑	☒	☒

Fig. 21. Evaluation matrix-wise analysis on the column scan techniques.

technique based BASIC_128, HYBRID_512 and HYBRID_1024 designs exponentially than BASIC_64 as the word size increases. As mentioned earlier, hybrid architecture merged two 128-bit words to make one 256-bit word. So, there are 2-bit *bit padding* in one 128-bit word for 3-bit column code. It extend to 4-bit *bit padding* for 256-bit word and so on. As a consequence, we are losing number of codes per word as the word size increases which effects the performance. On the other side, there is no chance of losing codes in *BitWeaving/V* as each bit of a code is store vertically per word (see Fig. 1(d)). This makes the marginal balance of processing codes per second between *Naïve/M* and *BitWeaving/V*. Therefore, performance-wise *Naïve/M* and *BitWeaving/V* both win over *BitWeaving/H*.

Technique-wise the behavior of throughput and performance are identical among the basic and hybrid architectures (see Fig. 18). Therefore, the speedup for main memory based intra-value parallelism based scan techniques among the basic and hybrid architectures on the targeted FPGA platform is linear (see Fig. 19), whereas the BASIC_64 design is the baseline. This defines, that the HYBRID_1024 design is best for all mentioned column scan techniques on FPGAs.

We also evaluated different numbers of bits per (column) code for three mentioned techniques using the best design: HYBRID_1024 (see Fig. 20). In this case symmetrical behavior found between all techniques, whereby a linearly decreasing behavior found for performance as the bits per code increases except the throughput. The reason is— increasing the code size decreases the number of codes per word which negatively effects the performance which is evaluated on the basis of the number of codes as expected, whereas throughput evaluation is independent of codes.

Table 2 illustrated the overall resource utilization in terms of LUTs (%) and FFs (%) for the best design HYBRID_1024 among all techniques using Xilinx® resource analyzer, whereby *Naïve/M* technique requires most optimum resource than the others due to its straight-forward predicate evaluation mechanism. After all kind of evaluations we found that, throughput-wise all techniques showed identical behaviour, performance-wise *Naïve/M* and *BitWeaving/V* techniques are better than *BitWeaving/H*, but resource utilization-wise *Naïve/M* technique is the most optimum one. Finally, these leads us to conclude that, *Naïve/M* technique is the best technique for FPGAs (see Fig. 21).

5   Related Work

Generally, the efficient utilization of SIMD instructions in database systems is a very active research field [17,25]. On the one hand, these instructions are frequently applied in lightweight data compression algorithms [5,12,24]. On the other hand, SIMD instructions are also used in other database operations like scans [7,23], aggregations [25] or joins [2].

Most research in the direction of FPGA optimization focused on creating custom hardware modules for different types of database query operations [10, 15,18,22,26]. For example, *Ziener et al.* presented concepts and implementations for hardware acceleration for almost all important operators appearing in SQL queries [26]. Moreover, *Sidler et al.* explored the benefits of specializing operators for the Intel Xeon+FPGA machine, where the FPGA has coherent access to the main memory through the QPI bus [18]. *Teubner et al.* performed XML projection on FPGAs and report on performance improvements of several factors [21].

Ever-increasing amount of data leads the recent research on main memory column store database system, whereas column stores are more effective performance-wise than row stores. In addition, it allows to evaluate query directly on the intra-data parallelism based compact storage layout. For that, there are several research has happened on how to efficiently evaluate query directly on the compact storage layout in order to improve the column scan performance, whereas the scan is one of the most important primitives in main memory database systems [7,13]. But to the best of our knowledge, none of the existing works investigated, firstly the domain of FPGA-accelerated data scan, secondly the comparison behavior as per intra-data parallelism based column scan mechanisms between FPGA-based and SIMD-based hardware implementation.

6   Conclusions

A key primitive in main memory column store database systems is *column scan* [7,13,23], because analytical queries usually compute aggregations over full or large parts of columns. Thus, the optimization of the scan primitive is very crucial [7,13,23]. In this paper, we explored two hardware-based implementation opportunities for scan optimization using SIMD extensions and custom architectures on FPGA on different scan mechanisms. In particular, we analysis the behavioral differences between *Naïve* [11] and *BitWeaving* [13] scan mechanisms as per hardware-based implementation. With both implementation, we are able to improve the scan performance, whereas the FPGA is best for *Naïve* technique and *BitWeaving* is perfect for SIMD. Therefore, to improve scan performance through FPGA do not require any fancy scan mechanism as *BitWeaving* due to its high parallelism criteria and flexibility to configure hardware as per requirements.

References

1. Abadi, D.J., Madden, S., Ferreira, M.: Integrating compression and execution in column-oriented database systems. In: Proceedings of the SIGMOD, pp. 671–682 (2006)
2. Balkesen, C., Alonso, G., Teubner, J., Özsu, M.T.: Multi-core, main-memory joins: sort vs. hash revisited. PVLDB **7**(1), 85–96 (2013)
3. Binnig, C., Hildenbrand, S., Färber, F.: Dictionary-based order-preserving string compression for main memory column stores. In: Proceedings of the SIGMOD, pp. 283–296 (2009)
4. Boncz, P.A., Kersten, M.L., Manegold, S.: Breaking the memory wall in monetdb. Commun. ACM **51**(12), 77–85 (2008)
5. Damme, P., Habich, D., Hildebrandt, J., Lehner, W.: Lightweight data compression algorithms: an experimental survey (experiments and analyses). In: Proceedings of the EDBT, pp. 72–83 (2017)
6. Faerber, F., Kemper, A., Larson, P., Levandoski, J.J., Neumann, T., Pavlo, A.: Main memory database systems. Found. Trends Databases **8**(1–2), 1–130 (2017)
7. Feng, Z., Lo, E., Kao, B., Xu, W.: ByteSlice: pushing the envelop of main memory data processing with a new storage layout. In: Proceedings of the SIGMOD, pp. 31–46 (2015)
8. He, J., Zhang, S., He, B.: In-cache query co-processing on coupled CPU-GPU architectures. PVLDB **8**(4), 329–340 (2014)
9. Hildebrandt, J., Habich, D., Damme, P., Lehner, W.: Compression-aware in-memory query processing: vision, system design and beyond. In: ADMS Workshop at VLDB, pp. 40–56 (2016)
10. István, Z., Sidler, D., Alonso, G.: Caribou: Intelligent distributed storage. PVLDB **10**(11), 1202–1213 (2017)
11. Lamport, L.: Multiple byte processing with full-word instructions. Commun. ACM **18**(8), 471–475 (1975)
12. Lemire, D., Boytsov, L.: Decoding billions of integers per second through vectorization. Softw. Pract. Exp. **45**(1), 1–29 (2015)
13. Li, Y., Patel, J.M.: BitWeaving: fast scans for main memory data processing. In: Proceedings of the SIGMOD, pp. 289–300 (2013)
14. Lisa, N.J., Ungethüm, A., Habich, D., Nguyen, T.D.A., Kumar, A., Lehner, W.: Column scan optimization by increasing intra-instruction parallelism. In: Proceedings of the DATA, pp. 344–353. SciTePress, Setúbal (2018)
15. Mueller, R., Teubner, J., Alonso, G.: Data processing on FPGAs. Proc. VLDB Endow. **2**(1), 910–921 (2009). 10.14778/1687627.1687730
16. Oukid, I., Booss, D., Lespinasse, A., Lehner, W., Willhalm, T., Gomes, G.: Memory management techniques for large-scale persistent-main-memory systems. PVLDB **10**(11), 1166–1177 (2017)
17. Polychroniou, O., Raghavan, A., Ross, K.A.: Rethinking SIMD vectorization for in-memory databases. In: Proceedings of the SIMD, pp. 1493–1508 (2015)
18. Sidler, D., István, Z., Owaida, M., Alonso, G.: Accelerating pattern matching queries in hybrid CPU-FPGA architectures. In: Proceedings of the SIGMOD, pp. 403–415 (2017)
19. Sidler, D., Istvan, Z., Owaida, M., Kara, K., Alonso, G.: doppioDB: a hardware accelerated database. In: Proceedings of the 2017 ACM International Conference on Management of Data, SIGMOD 2017, pp. 1659–1662. ACM, New York (2017). https://doi.org/10.1145/3035918.3058746. http://doi.acm.org/10.1145/3035918.3058746

20. Stonebraker, M., et al.: C-store: a column-oriented DBMS. In: Proceedings of the VLDB, pp. 553–564 (2005)
21. Teubner, J.: FPGAs for data processing: current state. IT Inf. Technol. **59**(3), 125–131 (2017). https://doi.org/10.1515/itit-2016-0046
22. Teubner, J., Woods, L.: Data Processing on FPGAs. Synthesis Lectures on Data Management. Morgan & Claypool Publishers, San Rafael (2013)
23. Willhalm, T., Popovici, N., Boshmaf, Y., Plattner, H., Zeier, A., Schaffner, J.: SIMD-scan: ultra fast in-memory table scan using on-chip vector processing units. VLDB **2**(1), 385–394 (2009)
24. Zhao, W.X., Zhang, X., Lemire, D., Shan, D., Nie, J., Yan, H., Wen, J.: A general SIMD-based approach to accelerating compression algorithms. ACM Trans. Inf. Syst. **33**(3), 1–28 (2015)
25. Zhou, J., Ross, K.A.: Implementing database operations using SIMD instructions. In: Proceedings of the SIGMOD, pp. 145–156 (2002)
26. Ziener, D., Bauer, F., Becher, A., Dennl, C., Meyer-Wegener, K., Schürfeld, U., et al.: FPGA-based dynamically reconfigurable SQL query processing. ACM Trans. Reconfig. Technol. Syst. **9**(4), 25:1–25:24 (2016)
27. Zukowski, M., Héman, S., Nes, N., Boncz, P.A.: Super-scalar RAM-CPU cache compression. In: Proceedings of the ICDE, p. 59 (2006)

Infectious Disease Prediction Modelling Using Synthetic Optimisation Approaches

Terence Fusco[✉], Yaxin Bi, Haiying Wang, and Fiona Browne

Ulster University, Shore Road, Newtownabbey BT37 0QB, UK
{fusco-t,y.bi,hy.wang,f.browne}@ulster.ac.uk
https://www.ulster.ac.uk/research/institutes/computer-science

Abstract. Research is presented in this work to improve classification performance when using real-world training data for forecasting disease prediction likelihood. Optimisation techniques currently available are capable of providing highly efficient and accurate results however, performance potential can often be restricted when dealing with limited training resources. A novel approach is proposed with this work known as Synthetic Instance Model Optimisation (SIMO) which introduces Sequential Model-based Algorithm Configuration (SMAC) optimisation combined with Synthetic Minority Over-sampling Technique (SMOTE) for improving optimised prediction modelling. The SIMO approach generates additional synthetic instances from a limited training sample while simultaneously aiming to increase best algorithm performance. Results provided yield a partial solution for improving optimum algorithm performance when handling sparse training resources. Using the SIMO approach, noticeable performance accuracy and f-measure improvements were achieved over standalone SMAC optimisation. Results showed significant improvement when comparing collective training data with SIMO instance optimisation including individual performance accuracy increases of up to 46% and a mean overall increase for the entire 240 configurations of 13.96% over standard SMAC optimisation.

Keywords: Over-sampling · Schistosomiasis · SIMO · SMOTE SMAC optimisation

1 Introduction

Optimisation techniques in machine learning aim to discover optimal parameter and model-based conditions for a given data set. Using an optimised approaches enables efficient removal of redundant or least effective algorithms from experiment conditions meaning improved testing processed for training data. Optimised solutions perform most effectively when using larger datasets due to speed capability for filtering vast amounts of data. These optimum processes involve

We would like to acknowledge The European Space Agency and The Academy of Opto-electronics, China as our research partnership in this work.

© Springer Nature Switzerland AG 2019
C. Quix and J. Bernardino (Eds.): DATA 2018, CCIS 862, pp. 141–159, 2019.
https://doi.org/10.1007/978-3-030-26636-3_7

using the most appropriate parameter variables or hyper-parameters in a training sample that enable the prediction model to best resolve the issue at hand. Hyper-parameter selection is an important tool in the optimisation process as it can concurrently target and improve weaknesses in a supplied data sample [2]. The context of proposed optimisation approaches in this research are related to disease prediction modelling for future prevention and control purposes. The specific problem this work is focused on is the epidemic disease known as schistosomiasis and the host vector freshwater snail. *Schistosomiasis* disease is caused by parasitic worms and infection takes place when freshwater that has been contaminated by the snail comes into contact with humans, crops and cattle resulting in a detrimental effect on those infected. According to the World Health Organisation, in 2015 around 218 million people required preventative treatment and over 65 million people were treated for infection in the same year. *Schistosomiasis* infections have increased in recent years in many parts of Africa and Asia meaning the need to provide early warning detection and prediction likelihood is imperative as a prevention and control tool.

This research aims to develop viable disease prediction models suitable for prevention and control measures to be implemented by relevant health bodies. Once this information is provided to communities at risk of disease outbreak, inhabitants can take known precautions to evade infection by avoiding unnecessary exposure to infested water bodies. The outcome of developing successful *schistosomiasis* prediction models for prevention and control purposes could drastically reduce the number of cases of people infected meaning reducing costs for treatment and adverse effects on livestock and crops. Authentic and plentiful real-world training data is of utmost importance when applying classification techniques for a particular disease research problem. A motivating factor for the proposed SIMO approach is the lack of training resources available for application of algorithms to construct disease prediction models. A solution is required that modifies the existing training sample in a way that can improve classification potential without noticeable change to original data in order to avoid undermining prediction results.

In contemporary related research, automated optimisation techniques provide significant performance improvements over standard applications leading to more technologically advanced approaches to many different research problems. Active learning approaches are becoming more prevalent in machine learning studies and are especially common with image classification research due to the nature of evolving discovery in that field. Optimisation approaches link both automation and active learning methods to find those features and parameters which perform most favorably for a specific data set [16]. The success of optimisation applications is often linked with larger datasets due to the ability to process information rapidly. Large, authentic training data can often be difficult to acquire in the epidemic disease exploration field, which has resulted in various techniques being constructed to amplify the sparse training data available. Some current sampling methods applied for improving imbalanced training data focus on using active learning and ensemble learning approaches, these have achieved good results by building on popular algorithms [10].

This research paper is an extension of a previously published work; namely [6] and further develops some of the concepts proposed in this publication. Specifically, initial SMOTE studies are included with Sect. 2 (2.3 SMOTE Equilibrium) including equations which describes the processing of imbalanced data application with SMOTE over-sampling method. Graphical representations are provided in Fig. 8 with results analysed and discussed in Sect. 4. Figure 7 depicts experiment results from proposed SIMO method which are also analysed and discussed with additional references added throughout the paper.

1.1 Contents

The structure of the rest of the paper is as follows; Sect. 2 will discuss some related work and notable research publications relevant to the current study subject area concerning data sampling methods. Section 3 covers the methodology of the proposed Synthetic Instance Model Optimisation (SIMO) approach along with some data sampling experiment results and a process diagram depicting how the method works. In Sect. 4, the results from all empirical experiments are provided with some analysis and discussion in review of the algorithm performance relative to the sample sizes applied. Section 5 will summarise the results gathered from experiments and provide conclusions while also highlighting significant results of this study.

2   Related Work

Optimisation techniques are becoming more popular for a variety of machine learning problems. Deep learning is one of the more recent areas of interest which involves optimisation techniques such as neural network training and optimised machine learning algorithms [1]. Improvement of experiment efficiency and performance enhancement are principle factors in the application of these methods for use in the context of epidemic disease forecasting. This focus on optimisation research can prove to be a vital tool for rapid information sharing pertaining to a variety of disease monitoring studies. Constraints of this research regarding sparse training data prompted investigation of sampling methods that could improve class balance and increase machine learning potential [14]. Optimisation and parallel algorithm simulations have been previously applied to physics research as well as epidemiology studies with success and in particular with the study of protein behaviors [19]. This *schistosomiasis* disease prediction research however, is more restrictive in terms of training data volume. Similar environment-based classification problems using sparse sample data can potentially benefit from findings in this work which face the same optimisation sample limitations. Opposing over-sampling and under-sampling techniques were considered and are assessed and expanded upon in this the following sections. Recent studies have compared real-world data and synthetic repository data for analysis of optimised active learning approaches which is a common method for optimised experiments [12]. Real-world data used in this research is used collectively and also as a base set for synthetic instance generation in order to assess the proposed optimised model in this study. Optimisation in machine learning requires an ever expanding number of training instances for comprehensive experiment conclusions and when this is not available, overall performance can be restricted. Sequential model-based optimisation is an approach which applies algorithms in an iterative sequential order to achieve optimum learning conditions [8]. Bayesian optimisation is another popular approach which employs an active learning procedure focusing on best performing algorithms during the optimisation process [5].

2.1 Data Sampling

Data sampling or re-sampling of skewed data is a common technique used in machine learning and specifically when using real-world data. Class imbalance issues can occur when using real-world environment data samples due to variations and density levels of spatial attributes. Over-sampling is an approach which uses additional sampling of instances in a supplied training set to increase the set size while balancing the data with increased minority classes. Under-sampling techniques are similar in that both methods share common aims but address the issue from different perspectives. Under-sampling is the inverse of over-sampling in that it reduces the size of a data sample by focusing on reduction of the majority class. A re-sampling approach was applied with this initial research to discover performance implications when using a limited training set. Similar conditions were applied to corresponding over-sampling experiments with the set used being the collective sample containing 223 instances and 8 attributes in total. The re-sampling method used provided a random sub-sample of the collective set using a bias to compensate for the minority class distribution. This was applied in fractions of the overall set for analysis with under-sampling of 100%, 80%, 60%, 40%, 20% and results recorded for analysis.

2.2 Synthetic Minority Over-sampling Technique

Synthetic Minority Over-sampling Technique (SMOTE) is a popular approach applied when using imbalanced sample data partly due to suitability for consecutive classification potential [3]. Increased training data can provide improved classification potential for optimisation application therefore, over-sampling techniques were deemed to be the most appropriate sampling choice for this sparse data problem. SMOTE is a sampling approach aimed at increasing a data set size with the purpose of improving minority class balance [15]. Synthetic instances are generated with minority bias as an alternative to over-sampling with replacement while also reducing the majority class hence increasing algorithm sensitivity to classifier assignment. For each instance x_i in the minority

Table 1. Raw data snapshot.

AREA	SD	TCB	TCG	TCW	MNDWI	NDMI	NDVI	NDWI
N49	0.03	0.26	0.13	−0.10	−0.58	0.03	0.60	−0.61
N60	0.02	0.29	0.10	−0.09	−0.45	−0.01	0.43	−0.45
N74	0.10	0.52	0.08	−0.07	−0.24	0.00	0.18	−0.24
N75	2.26	0.21	0.07	−0.03	−0.27	0.11	0.32	−0.37
N76	0.37	0.41	0.11	−0.15	−0.47	−0.09	0.32	−0.40
N77	0.08	0.21	0.13	−0.01	−0.30	0.29	0.56	−0.54

class, SMOTE searches the minority for the k nearest neighbors of x_i. One of these neighbors is selected as a seed sample $\hat{x}$. A random number between 0 and 1 denoted δ is chosen. The synthetic instance x_{new} is then created as shown in Eq. 1 [6].

$$x_{new} = x_i + (\hat{x} - x_i) \times \delta \tag{1}$$

2.3 SMOTE Equilibrium

This section is focused on data simulation and gives an insight into the validity of using a snapshot sample of environment data for disease vector classification, as opposed to the construction of an increased synthetic data set. For this reason, an over-sampling technique was deemed to be the most appropriate sampling choice for this data problem due to the increase in minority instance generation. Synthetic instances used in empirical experiments are generated based on original real-world training data. First of all the sample data was pre-processed then filtered before application of SMOTE to first increase the data to attain an equal number of classes, then incrementally increase the synthetic instance numbers to the required number. The SMOTE technique generates an increased number of synthetic data instances based on the original data set provided. SMOTE enables the construction of instances of data which aim to improve classifier performance by providing a balance of over-sampling the minority class and under-sampling the majority in a way that seeks to reduce the loss ratio during classification. The purpose of SMOTE application with this research is to address the issue of class imbalance and scarcity of data for training purposes. A research objective of this study is to assess the effectiveness of increasing training instances in terms of classification performance in contrast to smaller real-world samples. The SMOTE technique was implemented by pre-processing supplied raw field survey data then equalizing the number of SD classes from each year with synthetic instance generation should result in a larger training set with less over-fitting for applying classification methods.

The rationale behind SMOTE Equilibrium application is to utilize this approach with training data for comparative analysis purposes during the classification and prediction process. Ultimately, if it is the case that applying SMOTE can greatly improve classification performance over original data, then this process for optimisation can present a more bespoke model. The rationale behind proposing SMOTE is based on the fact that although there may potentially be access to future vast sources of satellite imagery to perform calculations for classification and prediction, this may not be the most computationally efficient approach or achieve the greatest performance. It is evident that a small training sample, as is used in this work, may not be representative of the greater population of SD environment data which is a fact that must also be considered during this process. Testing was conducted on each year of training samples in which the modified SMOTE method was implemented to achieve an equilibrium of SD classes to provide balance in the sample and eliminate the likelihood of over-fitting. This is a recurring issue with many labelling and classification research disciplines, which can skew results if not addressed. SMOTE has been applied to many studies in bio-informatics to deal with imbalanced datasets which suggests that this problem is a general issue in different areas of ML for classification which aims to avoid skewed experiment results. When using over-sampling as opposed to an under-sampling approach, the aim is to increase training potential of sample data set by providing a larger pool of data than the original limited set while increasing the minority class.

SMOTE Equilibrium Application was conducted with the initial analysis of SMOTE over-sampling resulting in a modified application of SMOTE with an equalled number of instance representations prior to applying the SMOTE increment increase. SMOTE over-sampling technique functions in a similar way however, to assess behaviour of equalizing the class distribution for each year in addition to other experiment results to assess the findings. This method was referred to as SMOTE Equilibrium due to the equalized class filtering prior to over-sampling application. SMOTE was initially applied to test the over-sampling approach and its capacity primarily to increase instance numbers while avoiding over-fitting of the data. This approach was implemented by taking each year of collected data and increasing the instances quantity to equalize the class numbers resulting in an equal number of SD classes before starting the training process. A simple calculation to show the process of the applied SMOTE Equilibrium instance increase is provided in Eqs. 2, 3 and 4.

$$diff = max - min \tag{2}$$

$$diff/min = inc \tag{3}$$

$$inc * 100 = inc \tag{4}$$

This method provided information on how the performance of the algorithms were changed with an increased number of training instances. The original training set was compared and used as a benchmark for analysis with the SMOTE Equilibrium method to record results when increasing from 100 instances in a training year up to 10,000 instances. The initial SMOTE study guided the next stage of research to test data simulation in terms of performance in comparison with real-world data and this helped to incorporate an optimised process into the model. Optimisation techniques are most appropriate for use when applied with a larger set of training data as an input as it can then run through a vast amount of iterations and provide a comprehensive set of test results. In this study however, a supplied sparse training set is used which has been synthesized from each year into a collective training set.

For this study results are evaluated using F-measure and classification accuracy performance metrics. F-measure calculations are shown in Eqs. 5, 6 and 7.

$$Precision = \frac{\sum True\ positive}{\sum True\ positive + \sum False\ Positive} \tag{5}$$

$$Recall = \frac{\sum True\ positive}{\sum True\ positive + \sum False\ Negative} \tag{6}$$

$$F - Measure = 2 \cdot \frac{precision \cdot recall}{precision + recall} \tag{7}$$

2.4 Sequential Model-Based Algorithm Configuration

Sequential Model-Based Algorithm Configuration (SMAC) optimisation is a method that seeks to optimise model parameters to the ideal setting before classifier application. SMAC optimisation is similar in many ways to Bayesian optimisation in that it also uses a sequential approach with active learning for providing optimised algorithm conditions [17]. It aims to find the best performing model and parameter settings for a particular data set in order to improve learning conditions for algorithms [18]. This is achieved using exploration of algorithm hyper-parameter space and includes examining new algorithms for performance analysis.

2.5 Research Issues

Issue that often arise when using automated optimisation processes are related to performance potential when using a sparse data set for learning. Sparse training resources can reduce effectiveness of optimisation capability and restrict potential for some automated techniques to be considered. The model proposed in this paper focuses on development of environment-based prediction models and issues surrounding the lack of real-world data for predicting vector-borne disease risk. Data samples used in this work provide evidence of class imbalance and this can be detrimental to the classification and prediction process [7]. Real-world data composition in general terms tends to contain unequally represented class categories. Training samples used encompass a six-year period from 2003–2009 around the Dongting Lake area in Hu'nan Province, China. From initial analysis and pre-experiment study phases, vector classes were identified that were unequally represented. Imbalance of this type can be due to a number of environment variables at the time of collection and is common with many real-world environment training samples. Over-fitting can occur when classifying imbalanced data due to a predominant class in the set rendering classifier tendency to assign that class label to new instances. An ideal solution for addressing class imbalance is to acquire additional data, which would increase the training pool

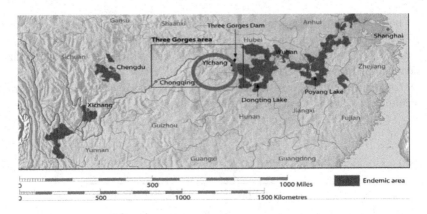

Fig. 1. Dongting Lake map.

and vary the class distribution. This is the most apparent approach when using a sparse data sample although difficulty lies in acquiring field survey information with corresponding freshwater snails. Lack of data and difficulty in accessing new samples is a significant component of the research problem being addressed with this work which is why the proposed method focuses on alternative generation of data for sample increase.

Developing disease prediction models can be restrained by limited field-survey data samples particularly in the case of vector-borne disease. Substantial collections of earth observation data have become more accessible in recent times however, corresponding vector distribution data can still be challenging to collect on a large scale therefore, optimisation approaches were investigated to maximize potential of classifier performance using limited data resources [4]. Motivation for this work focuses on providing early warning information to at-risk communities that can help with prevention and control of *schistosomiasis* outbreak and the destructive effects of transmission in local communities. Successful results and improved optimisation performance of proposed SIMO method will inform future research on optimal synthetic instance dimensions and model parameters for classification and prediction modelling. This can contribute to the advancement in optimisation and over-sampling approaches in any further experiment capacity in this epidemiology domain.

Modified forms of SMOTE over-sampling and SMAC optimisation currently exist and are useful tools for many research problems however, the proposed SIMO method provides a unified approach combining the two techniques in order to find optimum classifier performance which includes the optimum performing synthetic sample increase quantity. The research presented does not provide incontrovertible evidence of impending disease outbreak but rather the most informed advice for monitoring and control to present to health agencies dealing with public health risks.

Experiment training data used in this paper is supplied by research partners at the European Space Agency (ESA) in conjunction with the Academy of

Opto-electronics in Beijing, China. ESA partners provided satellite images over requested spatio-temporal parameters which was then used for environment feature extraction by Chinese partners at the Chinese Academy of Sciences (CAS). Feature extraction was conducted using spectral and spatial software for high resolution image processing which provided raw labeled environment values. This data was then presented and processed before being deemed experiment ready. The study area on which all experiments are based is the Dongting Lake area of Hu'nan Province, China as shown in Fig. 1 [6].

2.6   Training Data

Training data was provided using a combination of satellite information and environment feature extraction techniques which was then presented in a raw data format before preprocessing for experiment purposes. A snapshot of the training sample is provided in Table 1 [6]. Training data used in these experiments is a collective sample ranging from 2003–2009 containing 223 instances with eight attributes. The environment attributes used in all experiments are as follows:

- **TCB** - Tasselled Cap Brightness (soil)
- **TCG** - Tasselled Cap Greenness (vegetation)
- **TCW** - Tasselled Cap Wetness (soil and moisture)
- **MNDWI** - Modified Normalised Difference Water Index (Water Index)
- **NDMI** - Normalised Difference Moisture Index (soil moisture)
- **NDVI** - Normalised Difference Vegetation Index (green vegetation)
- **NDWI** - Normalised Difference Water Index (water index)

The theory and rationale for building the proposed prediction models is due to the fact that using satellite data and corresponding field-survey samples can help make informed prediction models for application with future satellite extracted environment data used for training accurate prediction models. The proposed synthetic optimsation method in this paper can assist in this research aim by assessing optimal classification parameters while evaluating synthetic instance generation viability on a sparse sample. The triumvirate of experts involved in this three-pronged research project are briefly described in Fig. 2 [6].

3   Methodology

To evaluate under-sampling of the sparse training set, a selection of established algorithms were applied having performed well in many fields of epidemic disease detection research to date [13]. These included Naive Bayes, J48, SVM and MLP and results are assessed using classification accuracy with corresponding sample size as presented in Fig. 3 [6]. Initial optimisation experiments were conducted on the collective training data to give the greatest data pool from the sparse samples for optimisation to take place. SMAC optimisation was applied

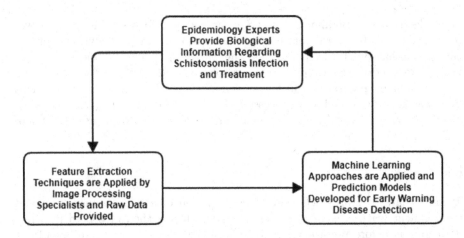

Fig. 2. Research partnership components.

with the top ten performing configurations being displayed for use in the next stage of testing [11]. Table 2 [6] shows experiment duration results ranging from 1–24 h of SMAC optimisation application with the collective training pool of 223 instances over a number of years from the Dongting Lake area of Hu'nan Province, China. At the end of each selected time period, the optimum performance algorithm together with weighted f-measure and classification accuracy findings were recorded. Each duration interval provided best performing algorithm results in terms of weighted f-measure and classification accuracy to provide a comprehensive algorithm analysis rather than accuracy metrics alone.

3.1 Synthetic Instance Model Optimisation

The proposed approach of this research is to implement Sequential Model-based Algorithm Configuration (SMAC) while simultaneously introducing an amplified number of synthetic instances using Synthetic Minority Over-Sampling Technique (SMOTE) to improve training potential with optimisation performance. In implementing this proposed Synthetic Instance Model Optimisation (SIMO), the aim is to increase performance of the optimised algorithm used to achieve greatest results. The success of this proposed method could alleviate the need to conduct much of the expensive and time-intensive field survey research that is required in order to make confident classification and prediction of the disease vector density and distribution. Results of this research will enable discovery of those classifiers which perform better with larger training sets of data and identify those poorly performing classifiers whose performance diminishes when increased synthetic instances are added. This information can be utilised for applying optimisation methods with future predictions.

Parallel algorithm configuration processing is a concept associated with optimisation and has been applied with success in the bioinformatics domain [9]. It is the assertion of this study that using optimised model processes in parallel with

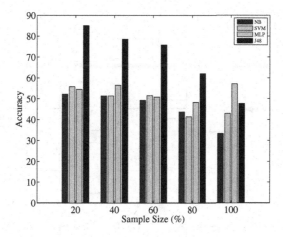

Fig. 3. Under-sampling results.

contributory sample balance improvement methods can significantly improve optimum performance potential of a sparse sample. The proposed SIMO model was constructed using a combined process involving SMOTE over-sampling and SMAC optimisation approaches. Combining approaches when optimising provides scope for improvement and can utilise the positive aspects of each respective technique. This SIMO approach is in essence an active learning approach which implements optimisation operations to both simulated instance sampling volume and model configuration selection.

The following stages describe the Synthetic Instance Model Optimisation (SIMO) process:

- SMAC optimisation is tested manually with the collective real-world training sample with duration intervals ranging from 1–24 h.
- SMAC optimisation is then applied in conjunction with generated SMOTE synthetic data simulation.
- The top ten best performing algorithms from each duration interval are then applied sequentially with synthetically generated instances for performance analysis.
- Both approaches are then unified into a single optimisation process with the objective of providing optimised synthetic instance generation models.

A model diagram of proposed SIMO approach is shown in Fig. 4 [6] showing the concurrent process which introduces training data to both SMOTE and SMAC techniques before beginning the unified SIMO approach. The experiment process involves running SMAC optimisation for every hour ranging from 1–24 h to assess performance of optimised techniques when applied with authentic sample data. Results of these initial tests were recorded and analysed for research purposes. Following this, the SIMO unified approach was applied with synthetically generated data based on the original sample ranging from 1000, 5000 and

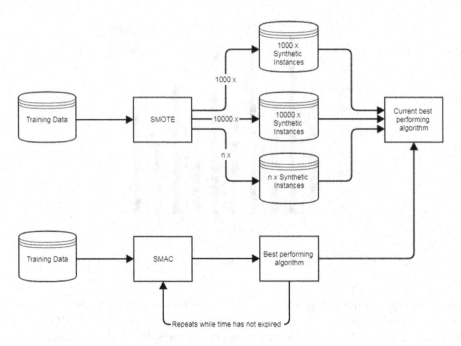

Fig. 4. SIMO process diagram.

10,000 instance gamut. In each of these synthetic sets, the first ten recorded optimised results were extracted and contrasted with the performance from the original data classification to assess effectiveness. During each of the experiment phases, a SMOTE Equilibrium approach was applied with increasing sample magnitude to appropriately assess the effects of the proposed synthetic data simulation approach.

4   Results

From initial results applied with an under-sampling technique in Fig. 3 [6], a gradual decline in performance accuracy it is noticeable with increases from 20% to 100% of the full sample size when under-sampling bias is implemented. This was expected from under-sampling of an already limited data pool but nonetheless contributed information of interest for assessing classifier behaviour with each batch increase. The performance of J48 decision tree decreased most significantly in terms of accuracy while MLP provided a performance gain between 20% and 100% sample size which can be factored into any future experiment thought process.

Figure 8 provides a number of line graphs to highlight the performance metrics of each algorithm over the 6-year range with number of instances added ranging from 1–10000. It is clear that the J48 classifier provides highly accurate and consistent results across each year with the exception of 2007 while Naive

Table 2. Benchmark optimisation results.

NumHrs	Algorithm	WeightedF	Acc%
1	RandomTree	0.982	98.2
2	J48	0.663	69.1
3	Logistic	0.583	66.4
4	OneR	0.746	77.6
5	RandomTree	0.982	98.2
6	RandomTree	0.982	98.2
7	OneR	0.991	99.1
8	Logistic	0.991	99.1
9	Bagging	0.622	68.2
10	Adaboost	0.605	59.2
11	Vote	0.609	65.5
12	RandomTree	0.559	67.7
13	Logistic	0.59	66.8
14	OneR	0.62	69.1
15	Bagging	0.722	74.4
16	Logistic	0.66	71.3
17	RandomSubSpace	0.622	62.8
18	RandomSubSpace	0.599	62.8
19	RandomSubSpace	0.599	60.1
20	LWL	0.702	73.1
21	LWL	0.721	74.9
22	OneR	0.684	71.7
23	LMT	0.555	67.7
24	OneR	0.684	71.7

Bayes presents little performance change over the generated instance increases. These results provide valuable information for developing future models which incorporate synthetic instances for prediction purposes.

In relation to graphical representations in Fig. 5 [6], a selection of results are presented to show optimisation performance from novel SIMO method in comparison with collective training optimisation configurations. In each of the hourly configuration accuracy results, the original collective data sample is denoted using C with synthetic instance volume represented by S followed by instance number in 1000, 5000 and 10,000 gamuts. Figure 5 [6] shows classification accuracy improvements in the majority of cases with increased synthetic instance simulation signifying optimisation performance improvements which shows scalability potential of the sparse training set. Similarly in Fig. 6 [6], significant f-measure performance improvements are noticeable when increasing synthetic

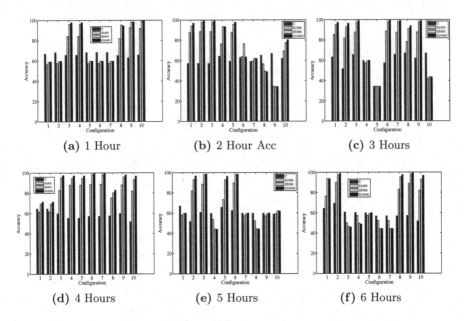

Fig. 5. SIMO accuracy 1–6.

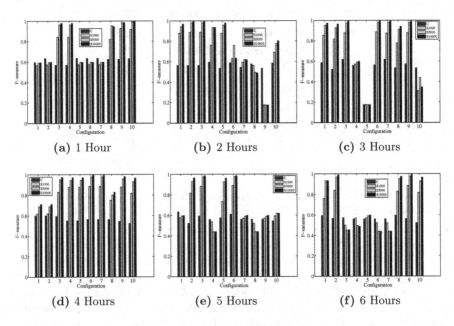

Fig. 6. SIMO f-measure results 1–6.

instances across the majority of testing models. These results using classification accuracy and f-measure metrics over a number of different optimised intervals, provide evidence to reinforce the necessity of proposed SIMO model as an effective tool for improving epidemic risk prediction modelling performance when using sparse sample data.

As an additional consideration, when analysing Figs. 7a, b [6], it is evident that with more synthetic instances added, accuracy prediction performance is reduced in many cases rather than improved as with the other results. These particular cases will require further investigation as to what characteristics of the data in that time range contribute to the performance levels or lack thereof.

4.1 Discussion

The method proposed in this paper is constructed using optimisation techniques in tandem with synthetic instance generation methods. The aim of this work is to find optimal conditions both in terms of parameter settings and instance simulation volume for making accurate classification of SD as well as discovery of environment attribute influence on SD levels. The predicted hypotheses of this work was that using the proposed SIMO method could improve accuracy and f-measure performance with synthetic instance optimisation over standalone SMAC optimisation during empirical experiments ranging in a 24-h temporal parameter setting.

The expectation from results using proposed SIMO approach is that performance improvement should be evident with each synthetic instance increment in comparison with optimised collective sample performance. This should be replicated with both accuracy and f-measure metrics in the main with some potential individual exceptions that will be identified for further analysis as is the case in Table 2 [6] in hour 3. In Figs. 5 and 6 [6], some of these results are presented affirming the initial prediction both in terms of accuracy and f-measure metrics.

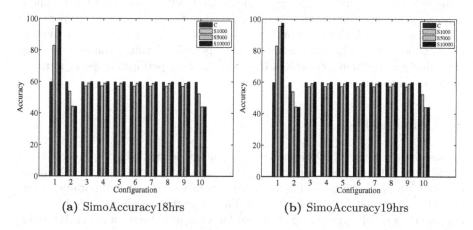

(a) SimoAccuracy18hrs   (b) SimoAccuracy19hrs

Fig. 7. Simo accuracy 1819 h.

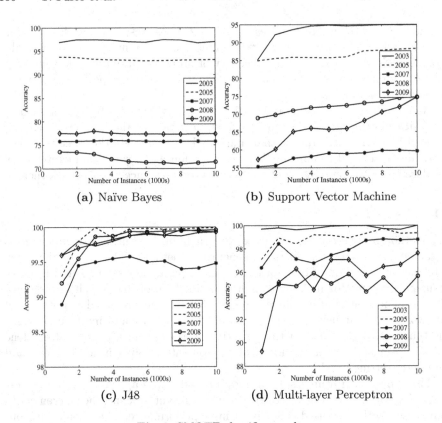

Fig. 8. SMOTE classifier results.

This validation has rendered the SIMO model an effective performance enhancing model suitable for use when applying optimisation approaches to a sparse training sample. There are however some anomalies in the results with a number of poor optimisation performances observed when applied over a longer period of time compared with shorter experiment duration's. These results require further investigation as to why performance was so poor in certain parameter settings and what the optimum classifiers from the poorest performing years were for future considerations.

5   Conclusions

In this study a novel SIMO method was presented using a hybrid approach incorporating SMAC optimisation and SMOTE instance generation with the aim of evaluating and assessing optimal instance generation volume and parameter settings for optimised classification. In summary, current findings have identified optimal parameter settings and classifiers for a range of duration intervals providing a knowledge base for future optimisation experiments in this field. Individual

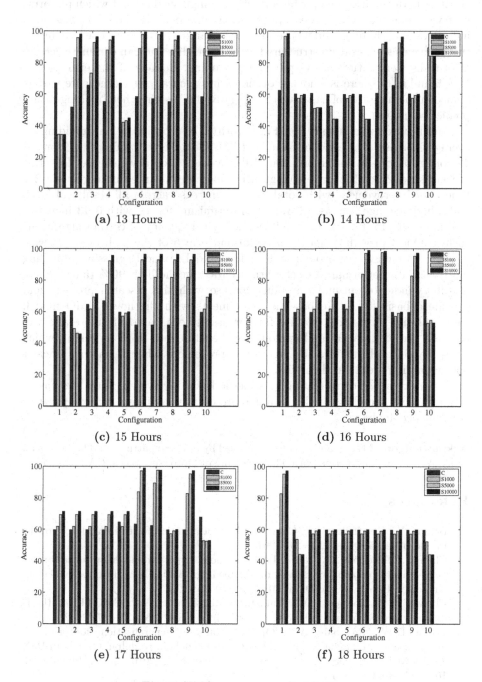

Fig. 9. SIMO accuracy results 13–18.

classifier performance can now be correctly distinguished as that which performs best with reduced or increased optimisation time periods. This information is indicative of each algorithm's potential for suitability with more machine intensive problems such as deep structured learning studies and can eliminate certain algorithms from future SIMO prediction training. In each of the examples in Figs. 5 and 6 [6], there is evidence of increasing accuracy and f-measure performance in the majority of cases which is positive for insight when building future predictive models.

Another example is shown in Fig. 9 with 13 h of optimisation providing an average increase of **27.2%** on standard SMAC implementation with 9 out of 24 configurations having more than **15%** average accuracy increase. In terms of f-measure, 25% of average configuration increases resulted in more than **2.5** f-measure improvement with total average increase across all results of **0.18** and a high average increase of **3.4** when optimising for 4 h with **0.43** increases in some cases. In Table 2 [6], results show high frequency of certain algorithms such as OneR providing optimum performance in 5 of the 24-h intervals with other similar regularity from Random Tree and Logistic Regression providing most accurate performance levels with accuracy in the high **90%** range. These classifiers indicate optimal suitability for use with this research problem and provide a basis for future baseline experiments with the novel SIMO model. The analysis factors that require further assessment based on all results will contrast the exploration and exploitation benefits that is, determining which performance level provides the greatest improvement while remaining efficient and maintaining data authenticity.

The next phase of validating this method will involve empirical evaluation of alternative sampling methods and representative datasets for comparative performance analysis.

Acknowledgements. Training data is supplied by research partners at The European Space Agency in conjunction with The Academy of Opto-electronics, CAS, China.

References

1. Bengio, Y., Goodfellow, I.J., Courville, A.: Optimization for training deep models. In: Deep Learning, pp. 238–290 (2015)
2. Chan, S., Treleaven, P., Capra, L.: Continuous hyperparameter optimization for large-scale recommender systems. In: Proceedings of 2013 IEEE International Conference on Big Data, Big Data 2013, pp. 350–358 (2013)
3. Chawla, N.V., Bowyer, K.W., Hall, L.O., Kegelmeyer, W.P.: SMOTE: synthetic minority over-sampling technique. J. Artif. Intell. Res. **16**, 321–357 (2002)
4. Corne, D.W., Reynolds, A.P.: Optimisation and generalisation: footprints in instance space. In: Schaefer, R., Cotta, C., Kołodziej, J., Rudolph, G. (eds.) PPSN 2010. LNCS, vol. 6238, pp. 22–31. Springer, Heidelberg (2010). https://doi.org/10.1007/978-3-642-15844-5_3
5. Feurer, M., Springenberg, J.T., Hutter, F.: Initializing Bayesian hyperparameter optimization via meta-learning. In: Proceedings of the 29th Conference on Artificial Intelligence (AAAI 2015), pp. 1128–1135 (2015)

6. Fusco, T., Bi, Y., Wang, H., Browne, F.: Synthetic optimisation techniques for epidemic disease prediction modelling. In: Proceedings of the 7th International Conference on Data Science, Technology and Applications, pp. 95–106. SCITEPRESS - Science and Technology Publications (2018)

7. He, H., Bai, Y., Garcia, E.A., Li, S.: ADASYN: adaptive synthetic sampling approach for imbalanced learning. In: Proceedings of the International Joint Conference on Neural Networks, pp. 1322–1328 (2008)

8. Hutter, F., Hoos, H.H., Leyton-Brown, K.: Sequential model-based optimization for general algorithm configuration. In: Coello, C.A.C. (ed.) LION 2011. LNCS, vol. 6683, pp. 507–523. Springer, Heidelberg (2011). https://doi.org/10.1007/978-3-642-25566-3_40

9. Hutter, F., Hoos, H.H., Leyton-Brown, K.: Parallel algorithm configuration. In: Hamadi, Y., Schoenauer, M. (eds.) LION 2012. LNCS, pp. 55–70. Springer, Heidelberg (2012). https://doi.org/10.1007/978-3-642-34413-8_5

10. Jian, C., Gao, J., Ao, Y.: A new sampling method for classifying imbalanced data based on support vector machine ensemble. Neurocomputing **193**, 115–122 (2016)

11. Kotthoff, L., Thornton, C., Hoos, H.H., Hutter, F., Leyton-Brown, K.: Auto-WEKA 2.0: automatic model selection and hyperparameter optimization in WEKA. J. Mach. Learn. Res. **17**, 1–5 (2016)

12. Krempl, G., Kottke, D., Lemaire, V.: Optimised probabilistic active learning (OPAL): for fast, non-myopic, cost-sensitive active classification. Mach. Learn. **100**(2–3), 449–476 (2015)

13. Lin, Y.L., Hsieh, J.G., Wu, H.K., Jeng, J.H.: Three-parameter sequential minimal optimization for support vector machines. Neurocomputing **74**(17), 3467–3475 (2011)

14. López, V., Triguero, I., Carmona, C.J., García, S., Herrera, F.: Addressing imbalanced classification with instance generation techniques: IPADE-ID. Neurocomputing **126**, 15–28 (2014)

15. Sáez, J.A., Luengo, J., Stefanowski, J., Herrera, F.: SMOTE-IPF: addressing the noisy and borderline examples problem in imbalanced classification by a re-sampling method with filtering. Inf. Sci. **291**(C), 184–203 (2015)

16. Settles, B.: Active Learning. Synth. Lect. Artif. Intell. Mach. Learn. **6**(1), 1–114 (2012). http://www.morganclaypool.com/

17. Snoek, J., Larochelle, H., Adams, R.P.: Practical Bayesian optimization of machine learning algorithms. In: Advances in Neural Information Processing Systems 25, pp. 1–9 (2012)

18. Thornton, C., et al.: Auto-WEKA: combined selection and hyperparameter optimization of classification algorithms. In: Proceedings of the 19th ACM SIGKDD International Conference on Knowledge Discovery and Data Mining, pp. 847–855 (2013)

19. Trebst, S., Troyer, M., Hansmann, U.H.E.: Optimized parallel tempering simulations of proteins. J. Chem. Phys. **124**(17), 174903 (2006)

Concept Recognition with Convolutional Neural Networks to Optimize Keyphrase Extraction

Andreas Waldis[ID], Luca Mazzola[(✉)][ID], and Michael Kaufmann[ID]

School of Information Technology,
Lucerne University of Applied Sciences, 6343 Rotkreuz, Switzerland
{andreas.waldis,luca.mazzola,m.kaufmann}@hslu.ch

Abstract. For knowledge management purposes, it would be useful to automatically classify and tag documents based on their content. Keyphrase extraction is one way of achieving this automatically by using statistical or semantic methods. Whereas corpus-index-based keyphrase extraction can extract relevant concepts for documents, the inverse document index grows exponentially with the number of words that candidate concepts can have. Document-based heuristics can solve this issue, but often result in keyphrases that are not concepts. To increase concept precision, or the percentage of extracted keyphrases that represent actual concepts, we contribute a method to filter keyphrases based on a pre–trained convolutional neural network (CNN). We tested CNNs containing vertical and horizontal filters to decide whether an n-gram (i.e, a consecutive sequence of N words) is a concept or not, from a training set with labeled examples. The classification training signal is derived from the Wikipedia corpus, assuming that an n-gram certainly represents a concept if a corresponding Wikipedia page title exists. The CNN input feature is the vector representation of each word, derived from a word embedding model; the output is the probability of an n-gram to represent a concept. Multiple configurations for vertical and horizontal filters are analyzed and optimised through a hyper-parameterization process. The results demonstrated concept precision for extracted keywords of between 60 and 80% on average. Consequently, by applying a CNN-based concept recognition filter, the concept precision of keyphrase extraction was significantly improved. For an optimal parameter configuration with an average of five extracted keyphrases per document, the concept precision could be increased from 0.65 to 0.8, meaning that on average, at least four out of five keyphrases extracted by our algorithm were actual concepts verified by Wikipedia titles.

Keywords: Natural language processing · Concept recognition ·
Convolutional neural networks ·
Keyphrase extraction Keyword extraction

1 Introduction

With the diffusion of user-generated content (UGC), the amount of information a knowledge management (KM) system should treat grows exponentially.

© Springer Nature Switzerland AG 2019
C. Quix and J. Bernardino (Eds.): DATA 2018, CCIS 862, pp. 160–188, 2019.
https://doi.org/10.1007/978-3-030-26636-3_8

On top of this, as the KM tool has no control over the source and format of the documents, it should be able to extract the major themes covered relying on some conceptualization process. To this end, the primary existing approach is called automatic document tagging and relies on the association amongst a series of keywords to each record based on their frequency and peculiarity for the document itself [28].

All-the-same, it is sometimes not advisable to rely on simple keywords, as they are unable to fully capture the semantic meaning of the subjects covered by an entry [4]. For this reason, we rely on the notion of concept, as a consecutive ordered sequence of single words. This is also known as n-gram, and the process for its identification as n-gram keyphrases extraction [26]. A concept is an idiomatic construction that conveys a meaning for humans used in this context for allowing them to have a feeling about the underlying subject(s). Building a comprehensive index of all the combinations of word stems for N-grams is a computationally complex problem because it grows over-linearly (combinatorially) with the increase of the length (n) of the sequence. Therefore, it is not feasible to tackle the problem in this way for a real case.

In this work, we measure the effects of applying a Convolutional Neural Network (CNN) as a filter on the set of the N-grams computed by a heuristic document based keyphrase extraction method with the objective to reduce the rate of false positives, that is, the percentage of keyphrases that are not concepts. The idea is to rely on the well-known capabilities of CNN to extract patterns from images [27] to use the set of extracted characteristics for identifying complex patterns in the construction of concepts as valid N-grams combination. A similar approach is already known in literature [22], where the objective is to identify matching in written text, treated as images.

The purpose of this research is to evaluate a neural network based algorithm to decide whether consecutive sequences of N words represent concepts. In this context, a concept is an idiomatic construction that conveys a meaning for humans. We use the English Wikipedia as labeled training corpus, We assume that all Wikipedia entry titles are concepts and our algorithm uses the existence of a Wikipedia entry for a given word combination as training signal. The resulting neural network should be able to recognize N-gram concepts in a given text. Those concepts can be used for entity extraction and automatic tagging without building a huge N-gram-based inverse document index. Such an algorithm that delivers proper n-gram concepts, regardless of the category and the size of the corpus, can increase the usefulness of keyphrases extraction for (semi–)automatic document classification or summarisation. In this work, we build upon a previously published conference paper [30], where the approach for concept recognition with convolutional neural networks has been described in detail. In the following sections describe our approach for keyphrase extraction and the application of CNN concept recognition as well as the measured effects based on data collected from experiments applying this new approach to a collection of Wikipedia articles. In Sect. 5 we summarize our insights from this research.

The rest of this paper is organised as follows: Sect. 2 explores the state of the art, in particular introducing the subjects of keyphrase extraction, automatic concept recognition, word embeddings, and the basics of convolutional neural networks. Most part of this section has been published by [30]. Section 3 is devoted to present concept recognition with CNN, from introducing our architecture, till evaluating its performance in isolation. Again, this Section is an excerpt from our conference paper [30], to make the basics clear for the reader. The new keyphrase extraction method with and without CNN is covered in Sect. 4, also using a toy demonstrative example. We present an experimental part by examining the performance impact of CNN–based filtering for keyphrase extraction based on collected data. This Section provides the core of our original contribution for this chapter. Some lessons learned and future steps that we would like to take conclude this paper in Sect. 5.

2   State of the Art

2.1   Keyphrase Extraction

Keyword extraction is the task of evaluating which words summarize the aboutness, or topics, of a text. However, often topics consist of more than one words, or n-grams. Keyphrase extraction aims at evaluating the central aspects of a text by ranking word groups regarding their relevance. Several approaches exist. Clustering-based methods such as Liu et al. [17] automatically group documents together and give clusters a label. Linguistic approaches such as [10] use grammatical structures based on part-of-speech tagging to generate keyphrases, and apply metrics based on statistical distribution to evaluate the relevance of keyphrases for the text. Graph-based methods [1] model documents as a graph, where nodes are words or phrases and edges are some form of relation between nodes, such as co-occurrences. For example, Text Rank [19] uses single words as nodes and co-occurrences in a window of two words as relationships. The importance of keyword candidates is then evaluated by a score based on the ratio between the sum of incoming edges weight divided by the sum of outgoing edges weight. In a post-processing phase, keyphrases are generated by joining direct sequences of n keywords in the text, to form n-grams. Corpus-based rankings such as TF-IDF evaluate the relevance of keywords and keyphrases based on the distribution of the symbols in the corpus [15]. On the other hand, document-bases approaches such as RAKE (Rapid Automatic Keyword Extraction) [25] operate only on distributions of candidate keywords in the document. RAKE computes more efficiently and achieves higher precision. The RAKE algorithm evaluates the relevance of keywords by dividing the frequency of a keyword candidate by the frequency of its co-occurrence with other keyword candidates. In a post-processing phase, keyphrases are extracted by joining keywords that occur at least twice in the same sequence, including the stop words between them. Keyphrase extraction methods usually focus on ranking keywords for relevance, and only apply simple heuristics or postprocessing to combine single words to keyphrases. Using concept recognition techniques, it is possible to generate more sophisticated methods to decide whether word combinations are concepts, which can improve keyphrase extraction.

2.2 Automated Concept Recognition

The importance of n-grams for text classification was shown already 20 years ago [7]. Many statistical and semantic methods have been proposed for concepts extraction. The use case described by [32] is an example for the usage of traditional neural networks, and [5] for the statistical approach. More recently, deep learning (an extension of neural networks with multiples hidden layers) is gaining relevance for all aspects of NLP, as mentioned by [18]. Concept recognition, as sub–field of concept mining, divides phrases into sequences of consecutive words classified as concepts and non-concepts. According to [23] concepts are useful by providing standalone information, in contrast to any random non-concepts. This information, as in [4], can be categorized as object, entity, event, or topic. For instance, the string *"the Guardian newspaper was founded in 1821"* contains 28 n-grams with the length of one to seven. The concept *"Guardian newspaper"* is one of them and has a significantly higher information level than the non-concept *"newspaper was founded in"*. There are several different approaches for deciding whether a phrase is a concept. [23] showed a combination of linguistic rules and statistical methods. The authors defined these rules to characterize possible concepts and filter out non-concepts. For example, a candidate has to contain a minimum of one noun and is not allowed to start or end with a verb, a conjunction, or a pronoun. After filtering out non-candidates, the remaining ones are judged by their relative confidence. This is a metric to help deciding if a sub-/super-concept of the candidate actually fits better as a concept. For example, *"Guardian newspaper"* is a better choice than *"Guardian newspaper was founded in 1821"* because of the higher relative confidence. Another method is shown in [16] with regards to Chinese bi-grams. Like [23] they combine statistical methods with linguistic rules as well, but in contrast, they first calculate the statistical metric and then filter out the results with linguistic rules. For measurement, they used the mutual information (*MI*) and *document frequency* (*DF*) metrics. *MI* represents the joint probability with respect to the product of the individual probabilities, for two words in a bi-gram. Since *MI* tends to prefer rare words, they used the *DF* value to reduce the influence of low-frequency words, as it takes into account the number of documents containing a bi-gram, normalized by the total number of documents.

2.3 Word Embeddings

Embedding $f : X \hookrightarrow Y$ map an object from a space X to another object of the space Y. One of the usages of embeddings in the field of NLP is, for example, to map a word (an item of the space of all words) to a vector in a high-dimensional space. Since these vectors have numerical nature, a wide range of algorithms can use them. The three mainly used embedding algorithms are *Word2Vec* [24], *GloVe* [31], and *fastText* [12]. While *GloVe* uses statistical information of a word,

Word2Vec and *fastText* adopt co-occurrence information to build a model. They calculate word embeddings based on either the continuous bag of words (*CBOW*) model or the *skip-gram* model of [20]. Those latter models predict respectively a word based on surrounding words (*CBOW*) or the surrounding words based on one word (*skip-gram*). *CBOW* and *skip-gram* rely on an input matrix (W_I) and an output matrix (W_O) as weight matrices. Those randomly initialized matrices are updated after each training iteration. The purpose of these matrices is to connect the neural network input layer to the hidden layer through W_I and the hidden layer to the output layer through W_O. In both methods W_I has the dimensions $V \times N$ and W_O the dimensions $N \times V$, where V represents the size of the vocabulary and N the size of the hidden layer. After optimizing these weight matrices, they can be used as a dictionary to obtain a vector for a specific word $h = x * W_I$, as discussed in [24].

Nevertheless, *Word2Vec* is only able to compute vectors for trained words, as it uses the vector of the whole word. One main advantage of *fastText* is the possibility of getting a word vector of an unknown word. To achieve this, it uses the vector's sum of sequences of included characters of one word, instead of one word as a whole. For example, *where*, enriched by the padding symbols < and >, is represented by <*wh, whe, her, ere,* and *er*>.

2.4 Convolutional Neural Networks

As a variation of neural networks (NNs), convolution neural networks (CNNs) are often used in computer vision for tasks such as image classification and object recognition. Usually, they adopt a matrix representation of an image as an input and a combination of different hidden layers to transform the input into a certain category or object. These layers are used to analyze specific aspects of the image or to reduce its dimensionality. Word embedding enables the numeric representation of words as vector, and the representation of n-grams as a matrix. Furthermore, they can serve as an input for CNNs. [9,13,14] have shown various use cases for the usage of CNNs in language modeling and text classification. They all rely on the word vectors as a matrix input. Inside the network, they combine different layers to analyze the input data and reduce the dimensionality. Two main processes are used in the network to learn: *Forward propagation* represents the calculation process throughout the whole NN to predict the output data for given input data. Each layer of the network takes the output of the previous layer and produces its updated output. The next layer uses this output as a new input. This process continues until it reaches the last layer. A majority of the layers use a weight matrix to process this transformation. This weight matrix controls the connection; that is, the strength between the input neurons and the output neurons. Finally, the update over the time of these weight matrices represents the learning process of the entire network. *Back propagation* allows the adaption of the neuron's connections weight based on the error between the output label and the resulting prediction. The metric *mean squared error* is shown in Eq. 1, with y_p as the predicted value and the truth as t_p:

$$E = \frac{1}{n} \sum_p (y_p - t_p)^2 \tag{1}$$

The goal of the *back propagation* process is to adjust the network's weights to minimize the difference $y_p - t_p$. This is done by propagating the error value layer by layer back through the network, calculating the partial derivative of the path from the output to every weight. Equation 2 shows how the error of a network can be back propagated to the weight $w_{b_3,c}$:

$$\frac{\partial E}{\partial w_{b_3,c}} = \frac{\partial E}{\partial c_{out}} * \frac{\partial c_{out}}{\partial c_{in}} * \frac{\partial c_{in}}{\partial w_{b_3,c}} \tag{2}$$

After distributing the output error over all weights, the actual learning takes place. Equation 3 shows that the new weight is the difference between the actual weight and the error multiplied by the learning rate:

$$w_{b_3,c} = w_{b_3,c} - (lr * \frac{\partial E}{\partial w_{b_3,c}}) \tag{3}$$

The kind of transformation and the connections inside the network are defined by the different kinds of layers used. The following layers are the mostly used ones: *Convolution layers* are used to analyze parts of or reduce the dimensionality of the input by applying a linear filter to the input matrix. This is done by iterating a kernel matrix (K) of the dimension $k_1 * k_2$ through the whole input matrix (I). The kernel matrix represents the weights and is updated through *backward propagation*. After applying the convolution operation to the input matrix, the bias value adds the possibility of moving the curve of the activation function in the x-direction and improve the prediction of the input data. Subsequently, the non-linear function adds some non-linearity. Without that, the output would be a linear combination of the input, and the network could only be as powerful as the linear regression. Doing that allows the network to learn functions with higher complexity than linear ones [21]. *Pooling layers* reduce the complexity and computation demands of the network, without involving any learning. The layer uses the given input matrix and creates a more compact representation of the matrix by summarizing it. It typically works with a $2 * 2$ window matrix iterating over the input matrix without overlapping. There are different kinds of *pooling layers* such as *max-pooling* or *average-pooling*. Using *max-pooling*, the highest value of the four cells serves as the representation of those cells instead of an average value. *Dropout layers* randomly ignore a percentage of the input neurons. This process only happens during the network training, for validation and prediction. As [29] showed, applying the dropout mechanism in a network increases the training duration but also increases the generality and prevents overfitting the network to the training set. The dropout procedure changes for each training sequence, as it is dependent on the input data. *Flatten layers* reduce the dimensionality of the input. For example, they convert a tri-dimensional input $(12 \times 4 \times 3)$ into a bi-dimensional ones (1×144). *Dense layers* are used to change the size of the given input vector. This dimensionality change is produced by connecting each row of the input vector to an element of the output vector. This linear transformation uses a weight matrix to control the strength of the connection between one input neuron and the output neurons. Like *convolution layers*, a bias value is added after the transformation, and an activation function adds some non-linearity. Dense layers are often used as the last layer of the network to reduce the dimensionality to usable dimension to get the prediction value, for example, with the *softmax* activation function, mostly used to get the predicted category in a classification task.

3   Concept Recognition with Convolutional Neural Networks

N-gram keyphrase extraction is a dimensionally hard problem. With increasing n, more and more of the extracted n-grams tend to not represent concepts, which makes the corresponding keyphrases noisy. Training a neural network to recognize whether an n-gram is a concept or noise could help filter the space of candidate n-grams for keyphrase ranking. To achieve this, the following method was implemented to filter out keyphrases using a trained neural network so that more keyphrases actually represent concepts. The *input* of the neural network is represented by a list of all n-grams extracted from the English Wikipedia corpus, with a maximal length of 7, encoded into a fixed 300-dimensional matrix by the word embedding model. The neural network is *trained* by using the set of Wikipedia page titles as the gold standard for deciding whether a sequence of words represents a concept: if an n-gram corresponds to a Wikipedia entry title, the training signal to the neural network is 1; else 0. The *output* of the neural network, for each n-gram, is a prediction of whether it represents a concept or not, together with the probability. The *goal* is to maximize the precision of the concept list, to obtain a high hit rate. The *objective* is to support automatic document tagging with n-gram concept extraction. *Evaluation* of the neural network's output success, again, uses Wikipedia page titles. If the neural network classifies a word sequence as a concept, then this is a true positive (TP) if there is a Wikipedia page with this title; otherwise, it is a false positive (FP). If the network classifies an n-gram as a non-concept, then this is a true negative (TN) if there is no Wikipedia entry with that name, or else it is a false negative (FN).

3.1   Neural Network Architecture

Figure 1 shows the NN network architecture. The *input matrix* is fixed to the dimensions 7×300 and contains the vector representation of the *n-grams*. Since the network needs a fixed input size, whenever $n < 7$, the matrix will be filled up with zero vectors. After specifying the input of the network, the *convolution layers* start analyzing the given data. For this purpose, two separate network paths are built for analyzing the data in the horizontal and vertical directions. Each of those two paths includes multiple *convolution layers* with different dimensions to gather different perspective of the data. All layers use a one-dimensional convolution layer that maps the two-dimensional inputs to a one-dimensional output.

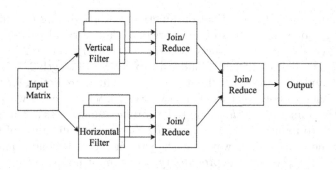

Fig. 1. Basic network structure (reprinted from [30]).

As shown in Fig. 2 the vertical *convolution layers* use filters with a fixed width of 300 and a dynamic height (here $2, 3, 4$). Correspondingly, the horizontal *convolution layers* are using a fixed height of 7 and a dynamic width (with typical values $30, 50, 70$). The *rectified linear unit* (ReLU), $f(x) = max(0, x)$, serves as

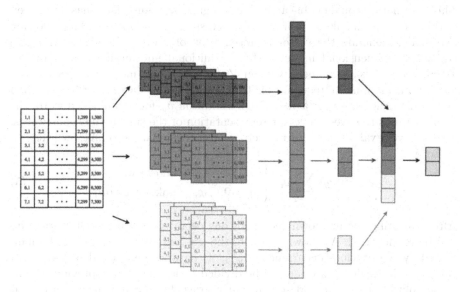

Fig. 2. Vertical convolution layers, with the different filter sizes (reprinted from [30]).

activation function for all filters, to add non-linearity to the output. Since the output dimension of the *convolution layers* is related to the filter size, the outputs of the filters are not balanced. For example, a vertical filter of size 2×300 produces an output vector of size 6 while an output vector for a filter size of 4×300 has length of 4. To deal with this aspect, after the *pooling layers* have reduced the complexity, some *dense layers* decrease the length of the output vector of each horizontal and vertical filter to 2, regularising them. They also use the ReLU function as their activation function. After reducing the dimensions, a *merge layer* joins all vectors of the horizontal and vertical paths. Then, the dimensionality of both resulting vectors will be reduced again to 2 by the usage of a *dense layer*. This eventually results in two vectors of length 2, 1 for the horizontal path and 1 for the vertical path. To get the final prediction for the input, a *join layer* merges both vectors of the two paths into one vector with length 4. Subsequently, a final *output layer* uses the *softmax* function as a dense layer (as seen in Eq. 4) to reduce the dimensions to 2. It squashes all values of the result vector between 0 and 1 in a way that the sum of these elements equals 1. The processed result represents the probability of one input n-gram being marked as *concept*.

$$\frac{exp(a_k(\bar{x}))}{\sum_j exp(a_j(\bar{x}))} \tag{4}$$

3.2   Word Embedding as N-Gram Features

For NLP applications, the choice of a *word embedding* plays a fundamental role, while also holding some contextual information about the surrounding words. This information could enable an NN to recognize n-grams that it has never seen in this sequence, thanks to a previously seen similar combination. The following two models generate the vector representation of a word: *Word2Vec* is a pretrained 300-dimensional model without additional information hosted by [8]. *Word2Vec-plus* is an extended version of the pre-trained model from [8]. It uses words with a minimum frequency of 50, extracted from a data set with 5.5 million Wikipedia articles. To get a vector representation v_u for an unknown word, the approach uses the average vector representation of the surrounding four words, if they have a valid vector, or the zero vector, if they are also unknown (shown in Eq. 5).

$$v_u = avg\left(\sum_{\substack{i=-2 \\ i\neq 0}}^{2} v_i\right) \begin{cases} v_i, & w_i = \text{known word} \\ v_i = 0, & w_i = \text{unknown word} \end{cases} \tag{5}$$

After averaging the unknown vector for one occurrence, the overall average v_n will be recalculated. As shown in Eq. 6, the existing average v_{n-1} will be multiplied by the previous occurrences w_{n-1} of the word and added to the vector calculated in Eq. 5. This value will be divided by the number of previous occurrences plus 1, to get a updated overall average for the unknown word. This variant of the average calculation prevents large memory consumption for an expanding collections of vectors.

$$v_n = \frac{v_{n-1} * w_{n-1} + v_u}{w_{n-1} + 1} \tag{6}$$

3.3   Training and Experimental Setup

Different network configurations were tested to find the best model for classifying *concepts* and *non-concepts* in our test case, according to the conceptual model described in the previous section. Figure 3 shows the iterative training pipeline:

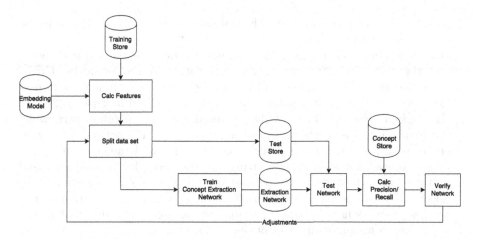

Fig. 3. The adopted training pipeline (reprinted from [30]).

1. Initially, the features of all n-grams found in the English Wikipedia corpus were calculated by extracting the corresponding word vectors from the *embedding model*.
2. Afterwards, the data set was separated into the training set (80%) and the test set (20%).
3. The training process used the training set to generate the *extraction network*. In this phase, the network was trained to recognize n-grams that are likely to represent a Wikipedia page title, based on their structure.
4. The prediction of the resulting network was based on the test set. During this process, all items of the test set were classified either as *concepts* or *non-concepts*.
5. The verification of the network was based on precision, recall, and *f1* metrics calculated during the previous evaluation.
6. To limit the human effort in classifying the results, we relied on the assumptions that valid concepts are statistically present as page names (titles) into Wikipedia, and that non-concepts are likely to not appear as page titles in this source.
7. Based on the performance comparison of a run with the previous ones, either the pipeline considered as finished or another round was started with an updated network structure.

The encoding used for presenting the data to the networks influences their performance, especially its ability for generalization. A good generalization depends mainly on the following three aspects:

– *Data balancing*: the input data is well-balanced by containing a similar amount of *concept* and *non-concept* examples. The final data set contains one million *concepts* and one million selected *non-concepts*. These two million samples do not fit completely into memory; thus dividing them into 40 parts avoids a memory overflow. The training process loads all of these parts, one by one for each epoch.

- *Data separating*: existing samples are separated into the *training part* (80%) and the *test part* (20%) to prevent overfitting of the networks. Cross-validation gives information about the level of generalization and mean performance. First of all, it separates the *training part* into four parts (25% of the original 80% set). Each of them is used once as *validation part*, while the remaining three parts serve as training data for the network. The precision, recall, and f1 score give a weight to each run of the cross-validation process of each network. Further changes to the network structure are based on these values to improve the performance. Also, those metrics are used to select the better performing and most stable networks. A final training run on these network uses the whole *training part* as input data and the *test part* to produce a final measurement of the best networks.
- *Shuffling* of the input data helps to get early convergence and to achieve better generalization, as also mentioned by [2]. For this purpose, the training environment loads all *training parts* in each epoch in a new random order. Furthermore, it shuffles all examples inside each part before generating the batches to send as network inputs.

Table 1 specifies the different network configurations adopted to investigate how vertical (v-filters) and horizontal (h-filters) filters can affect the performance of the resulting network. The intervals of the parameters shown in Table 2 are considered during the training. The actual combination differs by use case or experiment and is based on well-performing sets experienced during the whole project.

Table 1. Different model combinations (Source [30]).

Name	v-filters	h-filters
V3H0	(2, 3, 4)	()
V6H0	(2, 3, 4, 5, 6, 7)	()
V0H3	()	(100, 200, 300)
V3H1	(2, 3, 4)	(1)
V3H3	(2, 3, 4)	(100,200,300)
V6H1	(2, 3, 4, 5, 6, 7)	(1)
V6H3	(2, 3, 4, 5, 6, 7)	(100, 200, 300)
V6H6	(2, 3, 4, 5, 6, 7)	(10, 20, 30, 40, 50, 60)

Table 2. Hyperparametrisation (Source [30]).

Parameter	Value
Dropout	0.1–0.5
Learning Rate	0.0001, 0.0005
Epochs	100–400
Batch Size	32, 64, 128, 256

3.4   Evaluation of CNN Performance

Figure 4 shows the resulting *precision*, *recall*, and *f1* values in the training phase for all vertical and horizontal filter combinations, after the 400 training epochs. Here, some differences emerge:

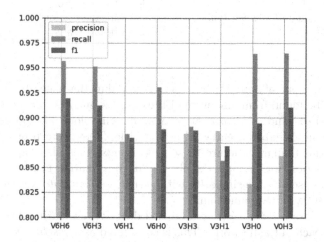

Fig. 4. Overall performance of the tested vertical and horizontal filters combinations in the training phase (reprinted from [30]).

- Based on *precision*, the V3H1 network had slightly better performance than V6H6 and V3H3, with a score of 0.8875.
- Considering the *recall*, on the other hand, the V0H3 (0.9625) architecture outperformed V6H3 and V6H6.
- Using the *f1* score, V0H3, V6H3, and V6H6 networks outperformed all others. However, among them, none has a significantly better performance, with all in the range from 0.91 to 0.9175.

Table 3 lists some classification examples, separated by their membership in the confusion matrix. *True positive* (TP) and *true negative* (TN) contain meaningful examples. The phrase *"carry out"* is an example of a *concept* that does

Table 3. Examples of neural network output, divided into true positives (TP), false positives (FP), false negatives (FN), and true negatives (TN) with regards to existing Wikipedia page titles. Objective is to minimise the FP set (Source [30]).

	Concepts		Non-Concepts
POSITIVE	American Educational Research Journal Tianjin Medical University carry out TP Bono and The Edge Sons of the San Joaquin Glastonbury Lake Village Earl of Darnley	FP	to the start of World War II must complete their just a small part a citizen of Afghanistan who itself include NFL and the a Sky
NEGATIVE	therefore it is use by FN in conversation with Council of the Isles of Scilly Xiahou Dun The Tenant of Wildfell Hall	TN	Regiment Hussars University of Theoretical Science Inland Aircraft Fuel Depot NHL and Mexican State Senate University of Ireland Station In process

not make sense out of context, but there is a Wikipedia page about it. Similar phrases can be found from among the FP examples, such as *"University of Theoretical Science"* and *"Mexican State Senate"*: they look like proper *concepts* but there is no Wikipedia entry with that title. They were probably selected because of the similar structure to some *concepts*. This also happened in the opposite direction; for example the phrase *"in conversation with"* is classified as *non-concept* based on the similarity with actual *non-concepts*; yet there is a TV series on BBC with the same name.

The test data set was restricted labeled data set, evaluating 20,000 selected n-grams for each of the networks. This is considered as a *baseline comprehension* measurement. For this purpose, the data set contains examples that have already been labeled by the different networks. This means the results produced by the existing solution were used as inputs. For each network the balanced data set contained, respectively, 5,000 *true positive*, *true negative*, *false positive*, and *false negative* examples. As the n-gram length can play a role in the performances, their total count and distribution between valid and non-valid concepts in the test data set are reported in Table 4. The unbalanced distribution with respect to this aspect is clearly evident, but this is reflected in the dataset characteristics.

An evaluation of each network using the test data set described in the previous section should give a feeling on how well they behave, on top of the statistical evaluation. For this purpose, separate data sets were used to compute the labels given by the NN. Eight data sets (one for each NN configuration) were initialized, each having a *precision* and *recall* of 0.5 as output of the corresponding NN. Table 5 shows the general performance of concept recognition in the test data sets. Figure 5 lists the resulting precision, recall and f1 for different n-gram lengths in the validation sets.

Table 4. Test data set distribution with respect to the n-gram length (Source [30]).

Length	Total count	Concepts	Non concepts
1-gram	90413	91.9%	8.1%
2-gram	164463	61.2%	38.8%
3-gram	107170	21.4%	75.9%
4-gram	52997	14.3%	85.7%
5-gram	20638	11.7%	88.3%
6-gram	8217	10.9%	89.1%
7-gram	3843	8.2%	91.8%

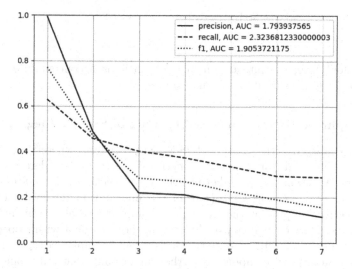

Fig. 5. Performances, with respect to n-gram length, of the CNN approaches, applied to a test dataset containing a list of 20,000 n-grams with an initial precision and recall output of the CNN of 0.5 for each CNN (reprinted from [30]).

In fact, the global performances of all networks negatively correlate with the n-gram length increases. Other than evaluating the whole test data set at once, the performance gaps of the different networks increased until some networks fell below 0.5. As before, V6H6, V6H3, and V0H3 outperformed their competitors; additionally, V3H0 performed almost equivalently. This suggests that they are more robust against unbalanced data and can achieve a more stable training process.

4   Keyphrase Extraction Method

With a CNN that recognizes concepts, it is possible to improve our keyphrase extraction (KPE) method. In this section, we will describe an initial statistical

Table 5. Performance results after application to a list of n-grams that have been tagged by different CNNs and have been balanced to represent 25% TP, FP, TN and FN examples (Source [30]).

Network	Precision	Recall	F1
V6H6	0.650	0.323	0.432
V6H3	0.659	0.326	0.436
V3H0	0.731	0.317	0.442
V0H3	0.694	0.331	0.448
V6H0	0.702	0.334	0.452
V3H3	0.649	0.348	0.453
V6H1	0.668	0.353	0.462
V3H1	0.640	0.366	0.466

approach for KPE, the integration of CNN to it as a concept filter into the process, and we measure its performance impact.

4.1   Keyphrase Extraction Based on Part of Speech Tagging

We initially started with a purely POS-based procedure, where the Wikipedia corpus[1] is analyzed for frequent functional patterns, to determine valid keyphrases. Additionally, we inherited the *term frequency* TF() function and *inverse document frequency* IDF() from the information retrieval filed. They each respectively represent the frequency of appearance of a term in a document and the relative frequency of the term in the document with respect of the full corpus. The underlying assumption is that valid complex concepts normally share the same type of components in their names and that only some specific patterns are useful for keyphrase extraction. We then measured the ratio for each POS pattern (eg: *(DT - JJ - NN - IN - NNP)*, instantiated by *the great wall of China*) between valid and invalid concept, such as in Eq. 7 and then weighted using TF-IDF to order them and extract the most significant ones, using the formula in Eq. 8.

$$\frac{P(pattern|C_+)}{P(pattern|C_-)} \tag{7}$$

$$\frac{DF(\boldsymbol{w})^2 + log(\frac{\#docs}{TF(\boldsymbol{w})+1} + 1)}{size(doc)} \tag{8}$$

The results demonstrated that this approach works very well with keywords (1–grams), but tend to eliminate keyphrases from the results. This is due two independent but concurrent phenomena: on one side, the part of speech filter removes many patterns including combinations as it tries to maximize the extraction precision; on the other side, the TF-IDF weight (in particular its Term

---

[1] https://dumps.wikimedia.org/enwiki/20181020.

Frequency part) is highly biased for very frequent elements, and elementary form are generally more frequent than composed ones.

4.2 Document-Based Keyphrase Extraction by Keywords Composition and Information–Based Weighting

Our keyphrase extraction approach is inspired by a human oriented idea, observing the way a person would search for significant keyphrases in a text. Through a visual scan, an individual searches for familiar patterns (as sequences of keywords, maybe also skipping a finite number of intermediate elements). This is important as keyphrases are often arranged around important keywords or are a combination of important words, as can thus be seen in our keyphrase extraction example in Fig. 7. We apply this idea in our method by using TF-IDF-based keywords as anchors to generate keyphrase candidates. We also combined the previously described POS-based filtering into the keyphrase extraction pipeline. By analyzing statistical characteristics of the document analyzed, the function $P()$ measures the probability of appearance of an element inside it. It can be used to compute individual probabilities, such as $P(w_1)$, or to compute joint probability of two words, such as in the case of $P(w_1, w_2)$. The first function we used is the *pointwise mutual information* [16], which computes the amount of information two words shares:

$$pmi(w_1, w_2) = \frac{P(w_1, w_2)}{P(w_1) * P(w_2)} \tag{9}$$

The limit of *pmi* resides on the impossibility to use it for comparing single words values with sequences. To solve this issue, we adopted another metric, based on the concept of the information amount included in a single word, called *self information* [11]

$$si(w_1) = \frac{1}{P(w_1)} \tag{10}$$

In order to compare the information gain for a words sequence a and its extension b (defined as a with an additional consecutive word), we defined a function called *average pointwise mutual information*, which represents for the given sequence the average pointwise mutual information between each pair of consecutive words.

$$avgPMI(\boldsymbol{w}) = \frac{1}{(n-1)} * \sum_{i=1}^{n-1} pmi(w_i, w_{i+1}) \tag{11}$$

Applying this functions allows us to answer the question: "How does the average information change if we add another word to the n–gram?", by comparing the avgPMI for a and b. Additionally, we defined the *mean probability* of an n–gram, as the weighted sum of each word probability inside the words sequence:

$$\bar{P}(\boldsymbol{w}) = \frac{1}{n} \sum_{i=1}^{n} P(w_i) \tag{12}$$

Furthermore to account for the positive co-occurrence of infrequent words, we defined the *Inverse standard deviation of each word probability in an n–gram,* which allows for higher weight to sequences which contains words with a similar probability:

$$invsd(\boldsymbol{w}) = \frac{1}{\frac{1}{n}\sum_{i=1}^{n}|P(w_i) - \bar{P}(\boldsymbol{w})|} \tag{13}$$

Eventually, the overall score combines the shown metrics into the following two formula for a generic sequence of words (including 1–grams (words) and n–grams):

$$score(\boldsymbol{w}) = \begin{cases} log(TF(\boldsymbol{w})) * si(\boldsymbol{w}) & \boldsymbol{w} \text{ is a proper word} \\ log(TF(\boldsymbol{w})) * (avgPMI(\boldsymbol{w}) + invsd(\boldsymbol{w})) & \boldsymbol{w} \text{ is a proper n–gram} \end{cases} \tag{14}$$

Given these functions, we defined the improved Keyphrase Extraction Method mode as follows. This process is illustrated in Fig. 6, where the pipeline from document to candidates tags is explained with reference to the following enumeration:

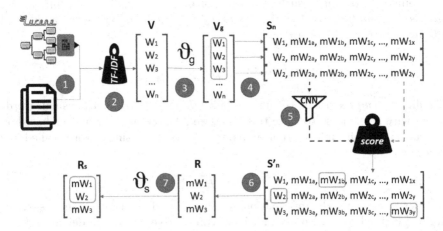

Fig. 6. The process flow of our proposed approach: 1-Grams are represented by W, and multigrams are represented by mW. The alternative paths with and without CNN usage are depicted as dashed arrows, respectively in black and yellow. Green highlighting is used to show the selected element in each step. (Color figure online)

1. We developed a 1-gram index on Wikipedia corpus and filtered it by part of speech as described in Subsect. 4.1 to create a keyword reference model, indicated as R.
2. Then, the approach found the most important words, in their singular form, based on their TF–IDF measure, and considered the elements in this ordered set V as keywords (1–grams).

3. A generation threshold θ_g limited the number of keywords to guarantee that a certain percentage of the information was conserved. For example $\theta_g = 0.8$ preserves around 80% of the information. This restricted set is called V_g.

4. On the original text we computed S as the set of all n–grams composed by maximum 7 elements contained in the respective document. For each element W in V_g we collected from S the candidates combinations containing W that were delimited by valid keywords. For example, given $W = States$ the combination *President of the United States of America* is a valid candidate, under the assumption $\{President, America\} \subseteq S$. Note that gaps are also allowed in valid combinations to allow inclusion of other POS elements such as prepositions, conjunctions, adverbs, and pronouns. This new multi set was named S_n.

5. The optional CNN concept recognition phase was applied at this level to improve the extracted keywords by filtering them for having a high probability of being concepts. In the following section, the quality of keyphrase extraction was evaluated with and without CNN-based concept recognition filtering.

6. A weighting procedure was adopted following the formula in Eq. 14. This allowed us to rank each candidate combination with respect to each element W. The one with the highest weight was then selected as representative. The resulting set was designated as S_n'

7. On the set R produced by the previous step, a dual filtering procedure was applied as the final stage based on the θ_s parameter: this criterion was applied on the original V_g, giving the best candidate for each initial keyword until the coverage of the expected information percentage. This allowed us to restrict

Database administrator_C.txt

Database administrator A database administrator (acronym DBA)
is an IT Professional responsible for the installation,
configuration, upgrading, administration, monitoring,
maintenance, and security of databases in an organization. The
role includes the development and design of database strategies,
system monitoring and improving database performance and
capacity, and planning for future expansion requirements. They
may also plan, co-ordinate and implement security measures to
safeguard the database. Skills List of skills required to become
database administrators are: Communication skills Knowledge of
database theory Knowledge of database design Knowledge about
the RDBMS itself, e.g. Oracle Database, IBM DB2, Microsoft SQL
Server, Adaptive Server Enterprise, MaxDB, PostgreSQL
Knowledge of Structured Query Language (SQL) e.g. SQL/PSM,

Fig. 7. Keyphrase extraction result for the Wikipedia entry "Database Administrator", as a screenshot in the UI of a prototypical software implementation.

the final set of suggested keyphrases, called R_s, in order to reduce to a set with higher information ratio.

As a first visual example, Fig. 7 shows the resulting keyphrases extracted for the Wikipedia article Database Administrator. In the text, the TF-IDF-based single keywords are highlighted. The goal of our method was to recognize frequent and important combinations of keywords and use them for composing n-gram keyphrases.

In a more detailed example, we will take a paragraph from a news website and present the proposed process using it. The website chosen is "the Washington Post" with an article titled "*The government is rolling out 2-factor authentication for federal agency dot-gov domains*"[2]. It is useful to note that this is not part of our corpus guaranteeing by construction the prevention of over-fitting effects. The selected paragraph is as follows:

```
Federal and state employees responsible for running government
websites will soon have to use two-factor authentication to access
their administrator accounts, adding a layer of security to prevent
intruders from taking over dot-gov domains.
```

Table 6. After step 2, extracted keywords are ordered by descending importance, together with their cumulative TF-IDF weight. This is the V vector. In step 3, using $\theta_g = 0.84$, the vector is limited to the top keywords with $\sum(weight) < \theta_g$, named V_g.

Keyword	weight	$\sum(weight)$	Keyword	weight	$\sum(weight)$
federal	0.130	0.130	employee	0.048	0.746
dot-gov	0.130	0.260	account	0.043	0.789
two-factor	0.116	0.376	security	0.043	0.832
authentication	0.087	0.464	*responsible*	0.042	0.874
intruder	0.076	0.540	*access*	0.042	0.916
administrator	0.055	0.595	*website*	0.041	0.957
domain	0.054	0.648	*government*	0.025	0.982
layer	0.050	0.698	*state*	0.018	1.000

As result of step 2, the set of limited extracted keywords V_g is presented in Table 6 together with their TF-IDF weight, rescaled to the unitary total, and its cumulated sum. In Step 3, the generation threshold θ_g is used to enforce the consideration of keywords from V till the cumulated TF-IDF measure do not exceed the parameter settled. Based on the limited set of keywords (V_g) the following potential concepts are found inside the text. N-grams with a length between 1 and 7 are allowed. Each valid concept has to start and end with one element of the limited set of keywords V_g, but inside the n-gram any word existing in the original text is allowed to appear. The result is shown in Table 7.

---

[2] The permanent link for the selected news item is https://perma.cc/PF53-SY2L.

Table 7. Step 4: Extracted candidate concepts for each keyword. This is indicated as S_n. The 1-grams corresponding to the keywords (first column) are not repeated.

Keyword	candidate Concept	candidate Concept
federal	federal and state employee	
dot-gov	dot-gov domain	
two-factor	two-factor authentication	
authentication	two-factor authentication	
intruder	security to prevent intruder	layer of security to prevent intruder
administrator	administrator account	
domain	dot-gov domain	
layer	layer of security	layer of security to prevent intruder
employee	federal and state employee	
account	administrator account	
security	layer of security	layer of security to prevent intruder

Step 5 is an optional part of the pipeline that preserves only the valid concepts by applying CNN filtering on the extracted candidates. In our example, using the best CNN configuration, this step eliminates the candidate concepts *"federal and state employee"*, *"security to prevent intruder"*, and *"layer of security to prevent intruder"* because they have a low probability of being a concept. From this point on, every candidate (also the 1-grams, previously known as keywords) are called concept; we proceeded with two cases, namely the result obtained with and without CNN filtering.

On step 6, the scoring of these candidates is performed following the formula in Eq. 14. This produces an order amongst the candidates for each keyword and then the highest scored concept is selected. Table 8 presents the selection of the set of concepts to use in vector R: grayed lines indicates elements filtered out by the CNN, as improbable concepts. The suit symbols indicates the selection of a specific concept to represent a keyword (in the columns): spade (♠) is used when the selection is common to both cases (with and without CNN filtering out), club (♣) pinpoints the top candidate without CNN filtering out, and heart (♡) is the highest remaining candidate whenever the previous concepts are removed by CNN filtering. Eventually, diamonds (◇) point to candidates that are not selected by either method.

Eventually, in step 7 the selection threshold is applied, to project back to the original cumulative information comparison. In our example, we use θ_s set to 0.72, that means in Table 6 to consider the top 8 elements (equivalent to the first column, till "layer", where the cumulated information is 0.698), that translates in Table 8 into the first 8 columns. This is the final result R_s, that is represented, by comparison, into Table 9.

Table 8. Step 6: The selection of the elements to compose the result R, based on the scoring function. Combinations marked by ♠ are selected in both cases (with and without CNN filtering); ♣ represents the top candidates without using CNN; ♡ indicates the highest remaining candidate after the CNN filtering step; ◇ represents candidates that are not selected by either method. The vertical separation refers to step 7, where a selection threshold $\theta_s = 0.74$ is applied, leaving only the first 8 columns in the resulting final set, know as R_s.

Concept	Score	federal	dot-gov	two-factor authentication	intruder	administrator	domain	layer	employee	account	security
federal	5.087	♡									
intruder	5.087				♡						
employee	5.087								♡		
federal and state employee	45.908	♣							♣		
dot-gov domain	45.908		♠				♠				
two-factor authentication	45.908			♠	♠						
security to prevent intruder	10.699				◇						◇
layer of security to prevent intruder	11.532				♣			♣			◇
administrator account	45.908					♠				♠	
layer of security	45.908							♡			♠

Table 9. The final candidate concepts for our small example comparing the case without (left) and with CNN (right). The number of candidates is not the same, as one concept can be associated to more than one keywords such as in the case of "layer of security to prevent intruder" in the removed concepts, here.

Concepts	
Without CNN	With CNN (best configuration)
federal and state employee	federal
dot-gov domain	dot-gov domain
two-factor authentication	two-factor authentication
layer of security to prevent intruder	intruder
administrator account	administrator account
	layer of security

4.3   Performance Impact of Adding CNN-Filter to Keyphrase Extraction

We can test the performance improvement by measuring the difference in precision, whenever we enhance the statistical keyphrase extraction with CNN-based concept recognition. If the CNN is applied as a filter, the n-gram keyphrase candidates can be improved in quality if the CNN can accurately predict concept-hood for n-grams. To test this, we implemented the following experimental setup. As a test set, we use 132.737 English Wikipedia articles of good quality: 18.487 FA (features articles), 27.738 GA (good articles) and 86.512 of quality level A. For every article, the keywords were extracted according to the process shown in Fig. 6 using different parameter configurations:

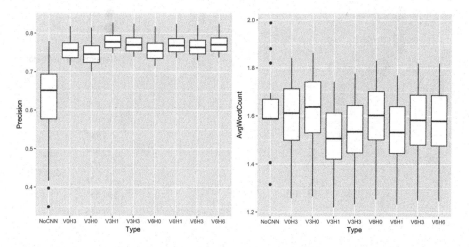

Fig. 8. Whisker-Boxplots of average precision and average keyphrase length (AvgWordCount) for every CNN configuration as discussed in Subsects. 3.3 and 3.4, compared to a configuration without CNN (the most left element, labeled as NoCNN).

- 10 levels of selection threshold θ_g: 0.48–0.84 in discrete steps of 0.04)
- 10 levels of generation threshold θ_s: again, 0.48–0.84 in discrete steps of 0.04, with the restriction of $\theta_s \leq \theta_g$
- 9 *Types* of keyphrase extraction, one without CNN and 8 with different CNN configurations, as described in Sect. 3.

For every Wikipedia article and for every parameter configuration $(\theta_s, \theta_g, Type)$, we collected the results as a list of extracted keyphrases. Now, for quality assessment, this list was compared to the list of concepts that have a Wikipedia entry. We assumed for our test that qualitatively good n-gram keyphrases are those that represent concepts, and if a keyphrase has a corresponding Wikipedia entry, we are certain that it represents a concept. Accordingly, a true positive (TP) is an extracted keyphrase that has a Wikipedia entry, and a false positive (FP) is a keyphrase without one. The precision of concept recognition in the list of extracted keyphrases for a document is then computed as the number of true positives divided by the number of keyphrases, meaning $\frac{\#TP}{(\#TP+\#FP)}$. We then summed up the document-level precisions to an average precision (macro-average) grouped by parameter combination. This allows us to compare the quality performance of different configurations. In the following paragraphs, we present the resulting data of these experiments.

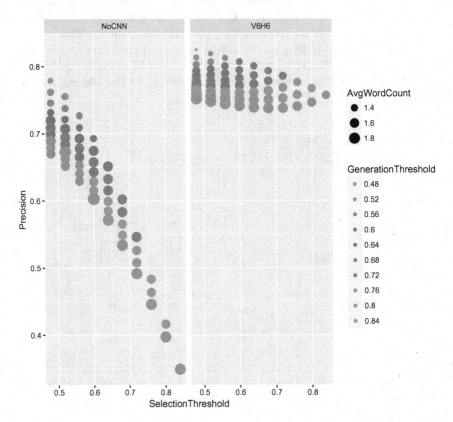

Fig. 9. Resulting precision and average keyphrase length (labeled AvgWordCount) without CNN (left) and with the optimal CNN configuration V3H1 (right), for different values of θ_g (color) and θ_s (independent axes). (Color figure online)

The left side of Fig. 8 shows that all CNN provide a significant improvement of concept precision compared to keyphrase extraction without CNN, and in particular the types V3H1, V3H3 and V6H6 showed the best improvements. Additionally, there are no more outliers (in particular towards the low end of the distribution) that is also a very good improvement. On the right side of Fig. 8 the average length (n) of the generated n-gram keyphrases is compared. As we can observe, some configurations of CNN tends to privilege shorter word composition as candidate keyphrases (such as V3H1, V3H3 and V6H1) whether others improves also on this aspect. In our use case of automatic tagging, precision is the most important criterion, in the sense of quality rather than quantity.

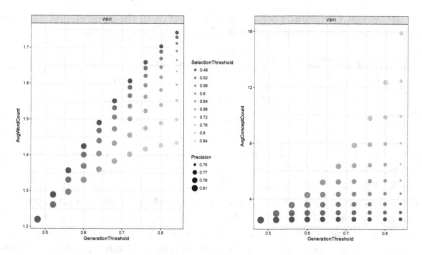

Fig. 10. Average number of words per keyphrase (AvgWordCount), average number of keyphrases per document (AvgConceptCount), and resulting precision (radius) of the optimal CNN configuration V3H1 for different values of θ_g (independent axes) and θ_s (grey scale).

Therefore, V3H1, the configuration that increases the precision the most, is considered as the optimal representative of CNN in our analysis, from this point on. We can now analyze the performance of our solution with respect to the different parameters used to tune it. The left side of Fig. 9 represents the concept precision achieved without CNN for different values of selection threshold θ_s that controls the amount of information preserved at the end of the pipeline, by selecting the top slice of keyphrase candidates. The right side of Fig. 9 compares this precision grouped by θ_s for the optimal CNN configuration V3H1. It is immediately evident that the base case (without CNN) is really sensible to increases in θ_s value because the precision drops as soon as you enlarge the pool as false positives more frequently appear in the candidate lists. On the contrary, the CNN seems to be largely immune from this effect, and it even appears to improve the precisions for larger values of θ_s. On the same graph, encoded by the point radius is presented the average length of generated keyphrase that show a very limited decrease of it, due to the introduction of the CNN in the pipeline. The color code represents the generation threshold θ_g, and what can be observed is that with a higher value of θ_g, the precision increases but also the average n-gram length (AvgWordCount) decreases.

In Fig. 10 the effects of the generation threshold θ_g on the length (left) and number (right) of candidates keyphrases is illustrated. This analysis is performed only for the pipeline including the CNN, in configuration V3H1. The color encodes the value of the selection threshold θ_s, whether the point size maps the achieved precision. An increase in θ_g produces, as expected, longer keyphrases, as more composed sequences emerges from the candidates; at the same time, the number of produced keyphrases positively correlate with the

Table 10. Result Comparison with and without CNN of the integrated pipeline with the same configuration. Notable is the increase in precision measure.

θ_s	θ_g	Type	Precision	AvgWordCount	AvgConceptCount
0.64	0.64	NoCNN	0.651	1.669	4.932
0.64	0.64	V3H1	0.806	1.360	5.190

selection threshold θ_s, as this causes more base keywords to be maintained, and they can be composed to define n-gram concepts. Also, the higher the selection threshold, the lower the average length of the extracted keyphrases.

If we assume an optimal number of five extracted keyphrases on average per document, we have an optimal configuration with parameters $\theta_s = \theta_g = 0.64$. As shown in Table 10, using this configuration we get a concept precision of 0.65 without CNN compared to a concept precision of 0.8 with a CNN with 3 horizontal and 1 vertical convolution layers, which is a clear improvement.

As an external evaluation, we measured precision and recall on the keyphrase extraction gold standard data sets Inspec and DUC according to the method described by Eiholzer [6]. First we tested for the optimal configuration. As shown on the right side of Fig. 11, the optimal CNN type of V3H1, consistent to the internal evaluation shown in Fig. 8. The left side shows an optimal θ_g parameter at 0.48 consistent to the internal evaluation shown on the right side of Fig. 9. Second, we compared our approach against data published by Bennani et al. [3]. Our approach with optimal configuration shows the highest precision (Table 11).

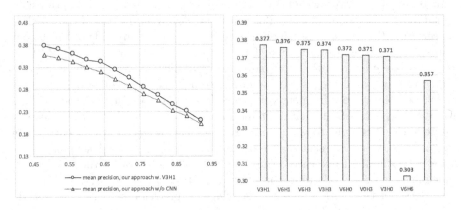

Fig. 11. The optimal configuration (left: $X = \theta_g$, right: $X = $ CNN type) according to the mean precision in an external evaluation of our approach on Inspec and DUC keyphrase extraction gold standard is $\theta_g = 0.48$ and CNN type = V3H1.

Table 11. Benchmarking of our approach ($\theta_g = 0.48$) on DUC and Inspec, on the *Precision, Recall* and *F1-measure* for top 5 keyphrase extraction. Approach$_A$ is without CNN and Approach$_B$ is with CNN in configuration V3H1.

Algorithm	Inspec			DUC			*Mean*		
	Precis.	Recall	F1	Precis.	Recall	F1	Precis.	Recall	F1
TextRank	0.249	0.105	0.147	0.198	0.123	0.152	0.224	0.114	0.150
SingleRank	.382	0.233	0.289	0.303	0.195	0.237	0.343	0.214	0.263
TopicRank	0.333	0.199	0.249	0.278	0.183	0.221	0.306	0.191	0.235
Multipartite	0.346	0.205	0.258	0.295	0.194	0.234	0.321	0.200	0.246
EmbedRank d2v	0.415	**0.254**	**0.315**	0.309	**0.197**	**0.240**	0.361	**0.225**	**0.277**
Approach$_A$	0.412	0.145	0.214	0.301	0.170	0.217	0.357	0.157	0.216
Approach$_B$	**0.442**	0.172	0.248	**0.313**	0.172	0.227	**0.377**	0.175	0.238

5   Conclusion

5.1   Insights and Lessons Learned

Our research contributes a soft computing method to recognize word sequences as concepts to filter the candidate concepts for key-phrase extraction. We trained a convolutional neural network using arrays of Word2Vec based embedding vectors (n-gram-matrices) as analytic signals and the existence of a Wikipedia title equal to the n-gram as a binary training signal to teach the CNN if an n-gram is a concept or not.

For the application of concept recognition to automatic tagging of documents, we are interested in high concept precision because false positives, that are keyphrases that are not concepts, decrease user acceptance of such a system. Also we are interested in n-gram concepts (with $n > 1$), as significant keyphrases usually consist of more than one word. Our experiments show that the concept recognition precision decreased with higher number of words in the n-grams, as from Fig. 5.

We contribute a statistical method for keyphrase extraction, and a solution to integrate CNN-based concept recognition to this keyphrase extraction to increase concept precision, the percentage of extracted keyphrases that are concepts. The experiments described in Subsect. 4.3 demonstrate that statistical keyphrase identification can be consistently improved using CNN-based concept recognition. We demonstrated an increase of concept precision for the extracted keyphrases calculated for Wikipedia articles, if extraction pipeline is combined with a CNN-based concept recognition filter. The data in Fig. 8 demonstrates that the percentage of automatically extracted keyphrases that are actual concepts was significantly enhanced by application of the CNN. Furthermore as seen in Fig. 9, whereas without CNN the average precision dropped quickly with increasing the number of keyphrases per document, this precision was more stable even with more extracted keyphrases adopting CNN filtering. Furthermore, as seen in Fig. 10, in a bench-marking experiment based on two key phrase extraction gold standard data sets, our approach with CNN-based concept recognition showed the highest precision compared to other methods found in the literature, even though it does not yield the highest recall. Since for automatic tagging, precision is more important than recall, because quality matters more than quantity, this result is optimal for our use case.

5.2   Outlook

There are several aspects to be considered for further research projects. Firstly, stemming from the gold standard could lead to a more general evaluation. Experimenting with different word features could increase the performance significantly. Secondly, we are currently experimenting with generating human-understandable features for words and n-grams based on co-occurrence statistics and explicit semantic analysis to generate a different type of word embedding. Thirdly we are comparing NN to Bayesian models and other types of models for concept recognition training towards getting a better understanding of the classification process. Fourthly instead of using a balanced list of concepts and non-concepts, the training data will be generated by going through the text corpus word by word. Thus, the network would be trained with n-grams in the sequence they appear in the text. Thus, frequent n-grams would be getting more weight. And last but not least, changing the input consideration and using a recurrent neural network instead of a CNN could improve the results.

Note

This is an extended version of a conference paper [30]. Equations (1), (2), (3), (4), (5) and (6) are reprinted with authors' permission from it.

Acknowledgements. This research has been funded in part by the Swiss Commission for Technology and Innovation (CTI) as part of the research project Feasibility Study X-MAS: Cross-Platform Mediation, Association and Search Engine, CTI-No. 26335.1 PFES-ES. We thank Benjamin Haymond for proof-reading and copy-editing of our work.

References

1. Beliga, S., Metrovic, A., Martinic-Ipsic, S.: An overview of graph-based keyword extraction methods and approaches. J. Inf. Organ. Sci. **39**, 1–20 (2015)
2. Bengio, Y.: Practical recommendations for gradient-based training of deep architectures. CoRR abs/1206.5533 (2012). http://arxiv.org/abs/1206.5533
3. Bennani-Smires, K., Musat, C., Jaggi, M., Hossmann, A., Baeriswyl, M.: EmbedRank: unsupervised keyphrase extraction using sentence embeddings. CoRR abs/1801.04470 (2018). http://arxiv.org/abs/1801.04470
4. Dalvi, N., et al.: A web of concepts. In: Proceedings of the Twenty-Eighth ACM SIGMOD-SIGACT-SIGART Symposium on Principles of Database Systems, PODS 2009, pp. 1–12. ACM, New York (2009). https://doi.org/10.1145/1559795. 1559797

5. Das, B., Pal, S., Mondal, S.K., Dalui, D., Shome, S.K.: Automatic keyword extraction from any text document using n-gram rigid collocation. Int. J. Soft Comput. Eng. (IJSCE) **3**(2), 238–242 (2013)
6. Eiholzer, M.: Method engineering for automatic tagging with inductive fuzzy classification. Master's thesis, School of Computer Science, Lucerne University of Applied Sciences and Arts, Rotkreuz, Switzerland (2019)
7. Fürnkranz, J.: A study using n-gram features for text categorization. Austrian Res. Inst. Artif. Intell. **3**(1998), 1–10 (1998)
8. Google: Googlenews-vectors-negative300.bin.gz (2013). https://drive.google.com/file/d/0B7XkCwpI5KDYNlNUTTlSS21pQmM/edit. Accessed 15 Jan 2018
9. Hughes, M., Li, I., Kotoulas, S., Suzumura, T.: Medical text classification using convolutional neural networks. arXiv preprint arXiv:1704.06841 (2017)
10. Hulth, A.: Improved automatic keyword extraction given more linguistic knowledge. In: Proceedings of the 2003 Conference on Empirical Methods in Natural Language Processing (2003)
11. Jagarlamudi, J., Pingali, P., Varma, V.: Query independent sentence scoring approach to DUC 2006. In: Proceeding of Document Understanding Conference (DUC) (2006)
12. Joulin, A., Grave, E., Bojanowski, P., Mikolov, T.: Bag of tricks for efficient text classification. arXiv preprint arXiv:1607.01759 (2016)
13. Kalchbrenner, N., Grefenstette, E., Blunsom, P.: A convolutional neural network for modelling sentences. CoRR abs/1404.2188 (2014). http://arxiv.org/abs/1404.2188
14. Kim, Y.: Convolutional neural networks for sentence classification. arXiv preprint arXiv:1408.5882 (2014)
15. Lee, S., Kim, H.: News keyword extraction for topic tracking. In: 2008 Fourth International Conference on Networked Computing and Advanced Information Management, vol. 2, pp. 554–559, September 2008. https://doi.org/10.1109/NCM.2008.199
16. Liu, Y., Shi, M., Li, C.: Domain ontology concept extraction method based on text. In: 2016 IEEE/ACIS 15th International Conference on Computer and Information Science (ICIS), pp. 1–5. IEEE (2016)
17. Liu, Z., Li, P., Zheng, Y., Sun, M.: Clustering to find exemplar terms for keyphrase extraction. In: Proceedings of the 2009 Conference on Empirical Methods in Natural Language Processing: Volume 1. Association for Computational Linguistics (2009)
18. Lopez, M.M., Kalita, J.: Deep learning applied to NLP. CoRR abs/1703.03091 (2017). http://arxiv.org/abs/1703.03091
19. Mihalcea, R., Tarau, P.: TextRank: bringing order into texts. In: Proceedings of the 2004 Conference on Empirical Methods in Natural Language Processing, EMNLP 2004 (2004)
20. Mikolov, T., Chen, K., Corrado, G., Dean, J.: Efficient estimation of word representations in vector space. arXiv preprint arXiv:1301.3781 (2013)
21. Nair, V., Hinton, G.E.: Rectified linear units improve restricted Boltzmann machines. In: Proceedings of the 27th International Conference on Machine Learning (ICML 2010), pp. 807–814 (2010)
22. Pang, L., Lan, Y., Guo, J., Xu, J., Wan, S., Cheng, X.: Text matching as image recognition. In: AAAI, pp. 2793–2799 (2016)
23. Parameswaran, A., Garcia-Molina, H., Rajaraman, A.: Towards the web of concepts: extracting concepts from large datasets. Proc. VLDB Endow. **3**(1–2), 566–577 (2010)

24. Rong, X.: word2vec parameter learning explained. arXiv preprint arXiv:1411.2738 (2014)
25. Rose, S., Engel, D., Cramer, N., Cowley, W.: Automatic keyword extraction from individual documents. In: Berry, M.W., Kogan, J. (eds.) Text Mining: Applications and Theory. Wiley, Hoboken (2010)
26. Siegfried, P., Waldis, A.: Automatische generierung plattformübergreifender wissensnetzwerken mit metadaten und volltextindexierung, July 2017. http://www.enterpriselab.ch/webabstracts/projekte/diplomarbeiten/2017/Siegfried.Waldis.2017.bda.html
27. Simonyan, K., Zisserman, A.: Very deep convolutional networks for large-scale image recognition. arXiv preprint arXiv:1409.1556 (2014)
28. Song, Y., et al.: Real-time automatic tag recommendation. In: Proceedings of the 31st Annual International ACM SIGIR Conference on Research and Development in Information Retrieval, pp. 515–522. ACM (2008)
29. Srivastava, N., Hinton, G., Krizhevsky, A., Sutskever, I., Salakhutdinov, R.: Dropout: a simple way to prevent neural networks from overfitting. J. Mach. Learn. Res. **15**(1), 1929–1958 (2014)
30. Waldis, A., Mazzola, L., Kaufmann, M.: Concept extraction with convolutional neural networks. In: Proceedings of the 7th International Conference on Data Science, Technology and Applications - Volume 1: DATA, pp. 118–129. INSTICC, SciTePress (2018). https://doi.org/10.5220/0006901201180129
31. Westphal, C., Pei, G.: Scalable routing via greedy embedding. In: INFOCOM 2009, pp. 2826–2830. IEEE (2009)
32. Zhang, Q., Wang, Y., Gong, Y., Huang, X.: Keyphrase extraction using deep recurrent neural networks on Twitter. In: EMNLP (2016)

Deep Neural Trading: Comparative Study with Feed Forward, Recurrent and Autoencoder Networks

Gianluca Moro[1]([✉]), Roberto Pasolini[1], Giacomo Domeniconi[2], and Vittorio Ghini[1]

[1] Department of Computer Science and Engineering, University of Bologna, Via dell'Università, 50, 47522 Cesena, Italy
{gianluca.moro,roberto.pasolini,vittorio.ghini}@unibo.it
[2] IBM TJ Watson Research Center, 1101 Kitchawan Road, Yorktown Heights, NY 10598, USA
giacomo.domeniconi1@ibm.com

Abstract. Algorithmic trading approaches based on news or social network posts claim to outperform classical methods that use only price time series and other economics values. However combining financial time series with news or posts, requires daily huge amount of relevant text which are impracticable to gather in real time, even because the online sources of news and social networks no longer allow unconditional massive download of data. These difficulties have renewed the interest in simpler methods based on financial time series. This work presents a wide experimental comparisons of the performance of 7 trading protocols applied to 27 component stocks of the Dow Jones Industrial Average (DJIA). The buy/sell trading actions are driven by the stock value predictions performed with 3 types of neural network architectures: feed forward, recurrent and autoencoder. Each architecture types in turn has been experimented with different sizes and hyperparameters over all the multivariate time series. The combinations of trading protocols with variants of the 3 neural network types have been in turn applied to time series, varying the input variables from 4 to 17 and the training period from 8 to 16 years while the test period from 1 to 2 years.

Keywords: Quantitative finance · Trading · Stock market prediction · Deep learning · Feed forward neural network · Recurrent neural network · LSTM · Autoencoder

1 Introduction

In financial economics, the theory of efficient-market hypothesis (EMH) [25] states that stock prices reflect all available information, including the future

This work was partially supported by the project "Toreador", funded by the European Union's Horizon 2020 research and innovation programme under grant agreement No 688797. We thank NVIDIA Corporation for the donated Titan GPU used in this work.
G. Domeniconi—Contribution done during the affiliation at the University of Bologna.

C. Quix and J. Bernardino (Eds.): DATA 2018, CCIS 862, pp. 189–209, 2019.
https://doi.org/10.1007/978-3-030-26636-3_9

expectations of traders and consequently the market can not be beat. This view, together with the result in [44], which highlighted adherence between stock values and the random walk theory, corroborate the unpredictability view of the stock market. Nevertheless, in [40] and in [43] the accordance between stock values and the random walk theory has been confirmed only within a short time window, namely the market trend appears generally predictable. The last two referenced works do not invalidated the EMH as the chance of predicting the stock price trends is not necessarily related to the market inefficiency, rather to the possibility of forecasting some macro economic variables.

Stock prediction methods, as extensively reported in Sect. 2, have evolved from auto-regressive ARIMA models [8] to advanced approaches based on machine learning, including neural networks in the last decade, using both structured and unstructured data, namely time series and free text from news, posts, forums and so on [4, 46, 50].

Novel proposals based on the combination of time series values with news or social network posts, such as in [3, 7, 18], claimed to overcome the accuracy predictions of preceding approaches. However combining both financial time series and news or posts to form input data, require daily large amount of fresh text which are almost impossible to gather in real time, even because the sources of news and social networks no longer permit unconditional massive download of their data. In other words the approaches based on this kind of unstructured data appear no longer viable for real time trading scenarios and this has renewed the interest on methods that require just the time series of stock values.

In the meanwhile machine and deep learning approaches have originated a large number of new results in several scientific and business areas, such as image recognition [35], diagnosis of diseases or bioinformatics [14, 15, 23, 38], astronomical discoveries, cross-domain text classification [16, 17, 20, 21, 48, 52], network sensors and security [10, 11, 42, 47] and so on. The breakthrough of deep learning solutions are due to new neural network architectures provided with memory capabilities. *Recurrent neural networks* (RNN) are the first memory-based learning technology that are able to self-detect long-term correlation patterns among the training instances, such as by analogy in sequences of words in sentences [19] or among consecutive values in historical series.

Surprisingly few studies so far have analyzed and supplied solutions based on RNNs for stock market trading. We think that one of the main reasons about the scarse adoption of RNNs for stock market trading, is that such neural networks are much more difficult to configure and train with success than traditional learning approaches. In fact the set of hyperparameters to tune an effective RNN, are mutually dependent and much larger than the past learning methods. Practically the configuration choices range from the architecture design, such as the neuronal units in each layer, the number of layers, the kind and number of neural connections among layers, the type of activation functions in each neuron unit, the learning rate and decay, the gradient optimization method, batch sizes, epochs, regularizations for feed forward and recurrent links and so on.

This work performs an entire new large set of experiments with respect to those reported in [24] where they were (i) focused on a single trading protocol, (ii) on two kinds of neural networks (iii) applied to the single time series values of the Dow Jones Industrial Average (DJIA). The experiments conducted with this paper combine further different neural network architectures and 7 trading protocols in order to compare their prediction and trading capabilities over 27 stocks. The selected multivariate stock time series, which are part of the DJIA, are composed by standard open, high, low, close values (OHLC) and by further financial variables. The experiments have been designed to also compare the performance of two subsets of such variables and of the full set.

The actions of the experimented trading protocols are driven by the neural networks predictions of the stocks' closing price movements. The trading protocols differ each other in the strategies of performing buy and sell operations.

The study is organized as follows. Section 2 analyses the recent literature for stock market predictions, from classical methods to those based on news and on social network posts. Section 3 illustrates our methodology with the data preparation, the composite neural autoencoders and the trading protocols. Section 4 describes the datasets, the neural networks' configurations and reports the experiment results. Section 5 summarises the work and outlines possible developments and improvements.

2   Related Works

In [36], in contrast to the unpredictability of the stock market, the authors explain the existence of a temporal lag between the issuance of new public information and the decisions taken by the traders. In this short time frame the new information can be used to anticipate the market. Following this idea that new information may quickly influence the trading, [54] combined the past stock prices with financial news through SVM to estimate the stock price 20 min after the release of new information.

Successively other works proposed methods that combine text mining techniques with usual data mining methods. [39] combines K-means and hierarchical clustering to analyze financial reports, which contain stock market operations, through quantitative and qualitative features, such as formal record of activities and position of businesses, persons and other entities [60]. This approach allows to detect clusters of data from such features which have been used to improves the quality of patterns in these reports that statistically lead to gains.

The most prominent approach used in time-series forecasting, which showed effective capabilities to short-term predictions, has been introduced in 1970 by Box and Jenkins and it is called ARIMA (Autoregressive integrated moving average [8]). ARIMA considers future values as linear combination of past values and past errors, expressed as follows:

$$Y_t = \phi_0 + \phi_1 Y_{t-1} + \phi_2 Y_{t-2} + \ldots \phi_p Y_{t-p} \cdots + \epsilon_t - \theta_1 \epsilon_{t-1} - \theta_2 \epsilon_{t-2} - \ldots \theta_q \epsilon_{t-q}$$

$$(1)$$

where Y_t is the actual stock value, ϵ_t is the random error at time t, ϕ_i and θ_i are the coefficients and p, q are integers called autoregressive and moving average. The main advantage of this approach, is the ability of observing it from two perspectives: statistical and artificial intelligence techniques. In particular in literature some studies validated the superiority of the ARIMA methodology with respect to artificial neural networks [45,57], but other works reported opposite outcomes [2]. Other works that used ARIMA are [32,33,37].

Behavioural finance and economy studies, such as [41], used cognitive psychology to recognize the logic behind the investor choices. This, together with the increasing computational power and the influence of the social networks, led to a new research thread focused to the correlation of sentiment acquired from online social opinions, such as social media or journalistic news, to the market trend. Some of these studies use mining technique to predict stock prices (e.g. the DJIA) [29,53], or to make more accurate predictions detecting relevant tweets such as in [18,49], exploiting also term weighting methods [22]. In addition, natural language understanding (NLU) and natural language processing (NLP) techniques together with deep neural network have been exploited to forecast the market trends [3,27,59]. The NLP has been used to extract latent information such sentiments or opinions from textual data and correlate them to the market price trends. With this approach, [58] found that media pessimism may affect both market prices and trading volume.

Similarly [7,18,49] enhanced the DJIA index prediction leveraging the mood from social media, like Twitter, extracted with text mining techniques; achieving a prediction accuracy of 86% and 88%, respectively, in a test set of one month. [7] also raised the attention on the *time-lag* of the used information. In their study, the best performance was achieved by grouping information from four past days, to predict the next one.

Also, the rising effectiveness of deep neural networks applied in the language understanding brought the trend also in the financial markets prediction. [3] exploited existing information regarding the companies, to predict stock values. For every day predictions, they used *Paragraph Vector* to create a distributed representation of news related to companies, feeding then these features into a memory network along with stock prices.

[13] introduced a novel dense representation of *event embeddings* from financial news. The approach leverage word embedding representation created from Reuters and Bloomberg news, that is then fed to a Neural Tensor Network (NTN) creating a distributed representation of event embeddings. The event embeddings are used with common prediction models to forecast the S&P 500 stock trend, achieving an accuracy of 65% and an economical profit of 67.85% with respect to the initial capital.

The use of NLP and NLU techniques are de facto inapplicable in real time scenarios, considering the massive amount of unstructured textual data that must be daily gathered and analyzed, mostly since the most popular social networks prevent unconditional massive data grabbing. In addition to other issues, such as how discriminating between relevant and irrelevant news, or how to behave in the majority of the days without news about a specific company, for which one should decide whether buying or selling its stocks.

However, several studies that have employed neural network to forecast stock market using only structured data, such as time series or financial data [5,55]. [51] tackled the annual stock returns prediction for 2352 Canadian companies. In this approach, 61 accounting reports are fed in a neural network, reporting better effectiveness than regression-based methods. Similarly [9] has shown that neural networks outperformed the accuracy predictions of linear models in Chinese stock market. The effectiveness of deep neural network in the prediction of stock returns in the cross-section is also reported by [1]. In their work standard machine learning models, shallow and deep network are compared in the prediction one-month-ahead of MSCI Japan Index stock returns.

In [26] LSTM networks have been compared with classical methods, among which *Random Forest*, for the prediction of the S&P 500 along a several years dataset (from 1993 to 2015). The obtained results were unstable in the three main phases in which the experiments were divided:

- 1993–2000: The LSTM outperformed the *Random Forest*, returning a cumulative profit of about 11 times the initial capital.
- 2001–2009: *Random Forest* performed better, obtaining an improvement of about 5.5 times the initial capital; while the LSTM cumulative profit was about 4 time the initial capital. This result was attributed to a greater robustness of decision tree-based methods against noise and outliers.
- 2010–2015: In this case LSTM network was able to reduce the prediction variability, keeping the initial capital unchanged, while *Random Forest* instead showed a loss of about 1.2 times the initial capital.

3   Methodology

3.1   Input Data and Predictions

As a reference for our method, we consider standard historical data derived from stock exchange, where for each market day we have the value of each stock indicated by *OHLC* values (*Open, High, Low, Close*). At the beginning of each market day t, other than historical data for all previous days $t-1$, $t-2$, ... we only have available the *Open* value for day t, hence $O(t)$. The goal we set is, for any given stock, to predict at the beginning of each market day t whether its *Close* value $C(t)$ will be greater or less than the corresponding *Open* value $O(t)$ on the same day. The general approach to attain this is to train a neural network to output a predicted close value $\hat{C}(t)$ based on information known at the beginning of market day t.

In the most basic setting we consider, such prediction is only based on knowing the *Open* value of the stock in both the current day and market days immediately before it. Several works have shown that there is a temporal lag between the issuance of new public information and the decisions taken by the traders. As shown by the experiments in [7,18,49] the higher correlation between social mood and the DJIA is obtained by grouping unstructured data of several days and shifting the prediction for a certain time lag. Similarly, [36] illustrated how the new information, publicly available, can be used within a short time window to anticipate the market. According with these approaches, we aggregate several days of information: considering a parameter d indicating the number of days to consider for prediction, prediction is based on values $O(t), O(t-1), \ldots, O(t-d+1)$.

In addition to the *Open* value itself, the prediction can also be based on the other available values, namely *Close*, *High* and *Low*. Differently from *Open* we can only consider sample of such values on market days strictly preceding t. Still considering a number d of days to aggregate, along with *Open* values $O(t), \ldots, O(t-d+1)$, we consider *Close* values $C(t-1), \ldots, C(t-d)$.

Along with OHLC values, we can additionally consider other daily-sampled indicators, including the trade volume (i.e. the number of traded stocks), trend-related technical indicators such as moving averages and macroeconomic variables.

In any case, to ease the training process, z-score normalization is applied to every variable given as input or target output to the neural network. For any variable X with sampled mean and standard deviation μ and σ, its normalization is given by

$$Z = \frac{X - \mu}{\sigma} \tag{2}$$

To properly replicate real operating conditions, the normalization of each value is only based on values available up to that time. For example, the z-score for a variable, say *Open* price, on 17/06/2009 is calculated considering only the prices available until 16/06/2009, leaving out those starting from 18/06/2009, respecting the conceptual integrity.

The next subsection introduces the architecture of the composite autoencoder that we experimented and compared in Sect. 4 with other two kinds of neural networks: feed forward and recurrent (RNN). Long Short-Term Memory (LSTM) [28,30,31] and Gated Recurrent Unit (GRU) [12] are the two most successful recurrent neural networks. The main difference between LSTM and GRU is the lack of explicit memory gates in the GRU networks and this reduces the number of tensor operations. Both solutions led to many recent state of the art results and although it is not clear which one is better, GRUs appear more efficient to train and generally more effective than LSTM on smaller training dataset. On the other hand, the higher complexity of LSTM architecture can make them able to correlate longer training sequences than the GRU networks. In this work where data should have complex sequence correlations we used LSTM.

3.2   Composite Autoencoders

The ability to extract meaningful features from raw input data is one of the key features of neural networks. *Autoencoders* are a class of neural network architectures which exploit such ability to perform tasks such as dimensionality reduction.

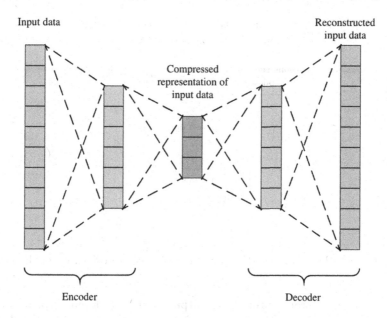

Fig. 1. General structure of autoencoders.

As depicted in Fig. 1, a typical autoencoder network is composed of two stacked groups of layers, an *encoder* and a *decoder*.

Such network is trained with a bunch of unlabeled data samples which are used both as inputs and as target outputs, so that the network is trained to replicate through the decoder output any information given in input to the encoder. By making its output size smaller than its input size, the encoder is trained to find a compressed representation of data which is as much loss less as possible, so that original data can be accurately reconstructed by the decoder. Once the network is trained, its decoding part can be discarded and the remaining encoder can be used as a stand-alone network to efficiently convert raw input data into the compressed representation.

This general architecture is usually implemented in feed-forward networks to be applied to fixed-size data with no notion of time. The encoding-decoding scheme can be conceptually applied also to time series, i.e. we can train a network to obtain a compressed representation of a sequence of values with possible correlations across different time steps: however, in practice, working with variable-length sequences and capturing such correlations is more complex.

Figure 2 depicts the autoencoder architecture based on LSTM layers we introduce in this work. A similar architecture has been proposed in [56] to learn representations of videos: layers of LSTM cells are used in both the encoder and the decoder. Authors run different tests on this architecture where the decoder network is trained either to reconstruct each frame in the sequence of the video given in input or even to predict the frames coming next in the sequence. In both cases, the encoder is first fed with all frames of the input sequence one

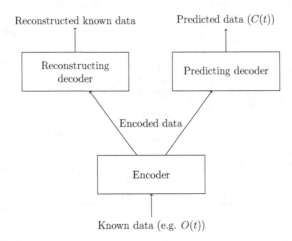

Fig. 2. General structure of the composite autoencoder defined for predictions.

by one, then the vector capturing its output at the last input step, whose size is thus independent from the length of the sequence, is picked by the decoder which gives in output the reconstructed frames or the predicted ones, still one by one.

For our time series prediction task, we considered specifically a composite model where the reconstruction of input sequence and the prediction of its continuation are performed by two parallel decoding blocks stacked on the same encoder. While we are actually only interested in the future values of a time sequence and not in reconstructing the present ones, by training the neural network on both tasks simultaneously we expect the shared encoder block to find a more precise and meaningful representation of the input sequence, thus improving the quality of the prediction.

3.3   Training of Neural Networks and Trading Protocols

We considered three possible architectures for the neural network used for prediction: a basic feed-forward network, a recurrent network based on two layers of LSTM cells and an autoencoder-based model with two decoders as presented in Sect. 3.2. The detailed structure of these networks is discussed later.

All networks receive as input the OHLC values of the target stock and possibly other values sampled on a daily basis. The *Open* value $O(t)$ presented for each market day t is relative to that day itself, while other values are relative to the previous market day $t - 1$. The target output for the network is the *Close* value for the current day. The predicted value $\hat{C}(t)$, compared with the known open value $O(t)$, indicates the suggestion given by the network to the trading agent.

$$\text{action} = \begin{cases} \text{buy} & \text{if } \hat{C}(t) > O(t) \\ \text{sell} & \text{otherwise} \end{cases} \tag{3}$$

In the training phase, one observation is generated for each market day of the considered training period. In the case of feedforward networks, inputs for d days going backwards from current day are presented at the same time, so that the network learns a combinatorial function of all such values. On the other side, for LSTM and autoencoder networks, input value are presented in a d-sized sequence to the network, so that it learns the general correlations across sequential market days.

After training the model for market prediction, in order to conduct an analysis of its real performance, we have to identify the economic gain in terms of dow points. We consider a trading agent which uses the model to decide the action to perform on each market day, which can be either *buy* or *sell* as illustrated in Eq. 3.

In our tests we consider multiple trading agents with different behaviors. Each of such behaviors is specified as a mix of traits indicating specific actions to be taken on each day conditionally to the decision indicated by the model. The following list describes the considered traits; the expressions *buy days* and *sell days* are used to refer to days where the model decision is *buy* and *sell* respectively.

- *Single purchase:* the agent buys a single stock on each *buy* day only if he doesn't already own one.
- *Daily purchase:* the agent buys a single stock on each *buy* day and may accumulate an arbitrary number of stocks.
- *Bulk purchase:* on each *buy* day the trader immediately acquires as much stocks as possible with the available capital.
- *Daily sale:* the agent sells a single stock on each *sell* day, provided he has at least one.
- *Bulk sale:* the agent sells all stocks he owns on each *sell* day.
- *Day trading:* on each *buy* day, the agent purchases a single stock (independently from how much he owns) at market day opening and sells it back at day closing.

All purchases of stocks are implicitly subject to having sufficient capital left. Also, any trade action (buying and selling) is always performed at the beginning of each day, thus considering the market opening value; the only exception to this is the selling action in the day trading trait.

All meaningful combinations of such traits are considered, leading to the following eight possible behaviors which will be hence identified by 2/3 letters initials:

1. **SD:** Single purchase + Daily sale
2. **SDD:** Single purchase + Daily sale + Day trading
3. **DD:** Daily purchase + Daily sale
4. **DDD:** Daily purchase + Daily sale + Day trading
5. **DB:** Daily purchase + Bulk sale
6. **DBD:** Daily purchase + Bulk sale + Day trading
7. **BD:** Bulk purchase + Daily sale
8. **BB:** Bulk purchase + Bulk sale

The resulting economic gain is then calculated as:

$$Gain = \frac{final\ capital}{initial\ capital} \tag{4}$$

The transaction costs resulted insignificant in the experiments and thus ignored in the analyses.

4   Experiment Settings and Results

Extensive tests were run to evaluate the combinations of the variants of 3 neural network architectures with the 7 trading protocols, in terms of accuracy of the predictions and of the expected profit.

4.1   Dataset, Training and Test Periods

We run a first batch of experiments on the selected stocks components of the Dow Jones Industrial Average (DJIA), including e.g. Apple, Microsoft, IBM etc. Table 1 provides a full list of the considered stocks. For each of them, we obtained historical data from Yahoo! Finance: we considered opening, closing, maximum and minimum daily values of dates ranging from 2000 to 2018. Each series was split in two reference periods:

- *training period:* from January 1st, 2000 to December 31st, 2015;
- *test period:* from January 1st, 2016 to November 30th, 2017.

Data from Yahoo! Finance only includes OHLC values and trade volumes. In order to also test our method on a richer set of input variables, we run a second batch of experiments on data used by [6]. Their dataset covers six stocks on a period ranging from July 2006 to September 2016. Other than OHLC values, their data includes several indicators typically used in technical analysis such as moving averages and also unrelated economical indicators; all such indicators are provided as additional input to the neural network. On this dataset, we consider the following reference periods:

- *training period:* from July 1st, 2006 to September 30th, 2014;
- *test period:* from October 1st, 2014 to September 30th, 2015.

Table 2 summarizes the variables involved in the prediction. The OHLC configuration is used for Yahoo! Finance datasets, while the complete configuration including all other variables is used for the stocks where their value was made available.

Table 1. Components of the DJIA index considered in trading experiments.

Symbol	Company	Industry
MMM	3M	Conglomerate
AXP	American Express	Financial
AAPL	Apple	IT
BA	Boeing	Aerospace/defense
CAT	Caterpillar	Construction/mining
CVX	Chevron	Oil & gas
CSCO	Cisco Systems	IT
KO	Coca-Cola	Food
XOM	ExxonMobil	Oil & gas
GS	Goldman Sachs	Financial
HD	The Home Depot	Retail
IBM	IBM	IT
INTC	Intel	IT
JNJ	Johnson & Johnson	Pharmaceuticals
JPM	JPMorgan Chase	Financial
MCD	McDonald's	Food
MRK	Merck & Company	Pharmaceuticals
MSFT	Microsoft	IT
NKE	Nike	Apparel
PFE	Pfizer	Pharmaceuticals
PG	Procter & Gamble	Consumer goods
TRV	Travelers	Insurance
UNH	UnitedHealth Group	Health care
UTX	United Technologies	Conglomerate
VZ	Verizon	Telecommunication
WMT	Walmart	Retail
DIS	Walt Disney	Entertainment

4.2   Design and Configurations of Neural Networks

We tested the proposed three neural network architectures: a plain densely-connected feed-forward network, a LSTM recurrent network and the LSTM-based autoencoder architecture. In the first two cases, each network is built with two hidden layers, with the second one having half the nodes (either neurons or LSTM cells) of the first; in the autoencoder architecture, the encoder is also based on two layers of LSTM cells with an half-sized second layer, while each of the two decoders contain one LSTM layer each with half the nodes of the second encoder layer. For the size of the first hidden layer used as reference for other

Table 2. List of variables used by neural network models.

Variable	Description
Target output	
$C(t)$	Closing value of current day
OHLC inputs	
$O(t)$	Opening value of current day
$C(t-1)$	Closing value of previous day
$High(t-1)$	Maximum peak of previous day
$Low(t-1)$	Minimum peak of previous day
Complete inputs (added to OHLC)	
All variables relative to previous day $(t-1)$	
Volume	Daily trading volume
MACD	Moving average convergence divergence
CCI	Commodity channel index
ATR	Average true range
BOLL	Bollinger band
EMA20	20 day exponential moving average
MA5/MA10	5/10 day moving average
MTM6/MTM12	6/12 month momentum
ROC	Price rate of change
SMI	Stochastic momentum index
WVAD	Williams' variable
Accumulation/distribution	
Exchange rate	US dollar index
Interest rate	Interbank offered rate

sizes we tested some different values. All layers use rectified linear unit (ReLU) as activation function.

During the training process, we use Adam [34] as the optimization method, with a base learning rate of 0.01. Training is performed in minibatches of 10 observations with a variable number of epochs. As stated previously, z-score normalization is applied incrementally to each input variable as it is given as input or as target output to the network. For each complete configuration of the neural network and of each training process, we executed 5 tests with varying random seeds, influencing the initialization of network weights and the distribution of training observations within batches.

All neural network architectures are implemented in Python using the Keras framework, TensorFlow is used as backend. We relied upon Keras default settings for any aspect not explicitly discussed above.

4.3   Trading Settings

After training a neural network on the whole training period, predicted close values are extracted for the selected test period and, by comparing them with real values, we obtain a vector of *buy/sell* decisions which are used in the trading experiments.

A trading agent with an initial capital of 50,000 dow points (USD) is considered for each of the trading protocols described above. According to decisions given by network predictions, each agent runs its own protocol for the whole testing period, forcibly selling any remaining stock on the last day of the experiment.

All following tables report the final capital owned by the trader at the end of the test period: for each considered test configuration the aggregation of 5 runs with different seed is shown in the form "mean ± standard deviation"; 1,000 dow points are used as unit.

4.4   Trading Tests on DJIA Components with OHLC Values

We present here the results on the first batch of tests performed considering the 27 single stock components of the Dow Jones Industrial Average (DJIA). In these tests we trained networks to formulate predictions upon OHLC values only. All these tests were performed using information aggregated from $d = 5$ days. Neural networks used were either FFN or LSTM with two layers containing 128 and 64 nodes or autoencoders with a base size of 32 nodes. 50 epochs of training were run an all networks.

Table 3 summarizes the results for each type of network and for each trading protocol, showing detailed values for trading experiments on the aggregated results on DJIA components. It also reports the accuracy of neural networks, intended as the percentage of predictions for which the sign of the variation from *Open* to *Close* values was guessed, independently from the specific values.

By comparing the three different neural network architectures, the more complex autoencoder network seems to effectively be more profitable with some trading protocols and yields results close to the others with the remaining protocols. Despite the measured accuracy is slightly lower, such network is presumably better at understanding and predicting general trends and providing accurate responses in market days where them matter most.

For what regards the trading protocols, we see first that some of them are more likely to cause a loss of capital rather than a gain. The problem seems to affect specifically the *daily sale* protocols where more than one stock may be owned but at most one is sold on each day: this may be due to the fact that traders of these types do not react quickly enough in the event of a relevant stock price drop lasting more days. On the other hand, the *bulk sale* protocols seem to perform best than the others, with BB being the one with the highest average gain across all tests. Due to their more conservative nature, *single purchase* protocols are less profitable than others, but the variance across different random executions is lower.

Table 3. Results obtained by trading DJIA components, grouped by neural network architectures and trading protocols.

Index	DJIA components (average)		
Model type	FFN	LSTM	AEnc
Avg accuracy	50.2%	50.5%	49.8%
Final capital (in 1,000s) by trading algorithm (initial capital = 50)			
SD	50.0 ± 0.0	50.0 ± 0.0	50.0 ± 0.0
SDD	50.0 ± 0.0	50.0 ± 0.0	50.0 ± 0.0
DD	32.9 ± 20.1	34.3 ± 19.9	41.2 ± 15.9
DDD	32.9 ± 20.1	34.3 ± 19.9	41.2 ± 15.9
DB	52.0 ± 5.1	51.4 ± 3.4	50.9 ± 3.2
DBD	52.0 ± 5.1	51.4 ± 3.4	50.9 ± 3.2
BD	9.5 ± 17.0	8.2 ± 16.3	12.4 ± 19.4
BB	57.9 ± 12.0	58.6 ± 12.0	56.2 ± 11.1

Being BB the most profitable trading protocol on an average, we report in Table 4 its results with the three different network architectures on each single DJIA component stock. Considering all average results over 5 runs, the best result is a final capital of 86,000 dow points achieved by the Autoencorder network with the CAT's stock: considering that the test period spanning almost two years, it corresponds to a yearly gain of 36% over the initial capital of 50,000 dow points. Considering that the overall average final capital by trading all the 27 stocks with the BB protocol and the LSTM is 58,600, the average annual gain in this case is 8.6%. In Table 4 five results overcome the threshold of 75%, which corresponds to 25% of annual gain.

4.5   Trading Tests with Additional Indicator Variables

The second batch of experiments whose results are reported here are based on the dataset of six stocks used by [6]: here the networks were trained using all variables listed in Table 2. Tests were run using aggregation over either 5 or 10 days. We also tested two different base sizes for each type of neural network: 128 and 512 for feedforward networks trained over 200 epochs, while 32 and 128 are the values tested for LSTM and autoencoder networks trained over 100 epochs.

Table 5 compares the performances of the different configurations using different trading protocols: those using the *daily sell* trait, not reported in the table for space reasons, cause losses of capital likely to tests of the first batch. Reported results are averaged over the six stocks used in experiments. It can be seen that differences across the tested configurations are minimum in terms of

Table 4. Detailed final capital (in 1,000s) obtained for each stock with the bulk purchase + bulk sell trading protocol (*BB*).

Stocks	FFN	LSTM	AEnc
AAPL	68.7 ± 16.8	77.8 ± 10.7	65.1 ± 15.9
AXP	59.3 ± 13.7	55.7 ± 7.8	52.1 ± 8.9
BA	75.9 ± 24.7	64.4 ± 21.6	60.8 ± 12.3
CAT	74.7 ± 18.4	76.9 ± 22.3	86.0 ± 25.6
CSCO	57.6 ± 7.9	60.2 ± 6.3	54.0 ± 6.2
CVX	59.3 ± 9.1	64.5 ± 8.9	56.9 ± 8.9
DIS	50.6 ± 0.2	49.8 ± 1.2	49.7 ± 5.0
GS	52.4 ± 5.1	59.5 ± 8.7	52.9 ± 3.2
HD	62.7 ± 10.1	57.8 ± 9.2	58.2 ± 8.8
IBM	53.1 ± 3.9	55.0 ± 3.2	54.3 ± 6.7
INTC	59.2 ± 6.7	57.0 ± 5.4	50.2 ± 4.7
JNJ	58.5 ± 9.1	61.3 ± 10.2	58.0 ± 6.6
JPM	54.3 ± 6.0	63.1 ± 16.6	49.8 ± 4.3
KO	52.5 ± 1.8	53.4 ± 1.9	54.2 ± 3.5
MCD	63.0 ± 11.9	61.4 ± 11.5	56.7 ± 9.6
MMM	63.9 ± 16.4	69.0 ± 18.0	54.8 ± 10.2
MRK	52.0 ± 3.0	59.4 ± 6.3	60.0 ± 4.5
MSFT	68.8 ± 11.2	57.1 ± 11.9	56.1 ± 6.5
NKE	50.5 ± 3.2	51.2 ± 4.3	53.4 ± 6.7
PFE	55.9 ± 4.2	56.7 ± 3.1	55.6 ± 3.9
PG	57.6 ± 0.4	56.0 ± 3.9	55.8 ± 8.6
TRV	57.2 ± 5.6	56.1 ± 5.0	54.1 ± 5.3
UNH	50.0 ± 0.0	50.0 ± 0.0	50.0 ± 0.0
UTX	58.8 ± 9.7	63.0 ± 7.6	56.6 ± 7.7
VZ	51.0 ± 4.3	49.9 ± 4.0	51.0 ± 3.6
WMT	64.5 ± 8.9	63.5 ± 7.3	66.8 ± 10.5
XOM	53.9 ± 3.1	55.4 ± 3.5	53.5 ± 4.6
Avg.	57.9 ± 12.0	58.6 ± 12.0	56.2 ± 11.1

resulting profit and that in most cases the aggregation of more days or the use of a larger network does not improve the final result. Also the differences between different trading protocols are much lower in these tests, partly due to the test period of only one year and possibly also to the shorter training period, leaving a smaller amount of training data.

Table 6 reports detailed results for each stocks considering the most lightweight configuration for each network type, i.e. aggregation of $d = 5$ days

Table 5. Aggregated results on the five stocks - CSI300, HangSeng, Nifty50, Nikkei225, S&P500 - predicting upon all variables for the input, comparing different configurations and trading protocols.

Model type	d	Model size	Model accuracy	Final capital in thous by trading algorithm, initial = 50				
				SD	SDD	DB	DBD	BB
FFN	5	128	49.1%	50.1 ± 1.1	49.8 ± 2.5	50.8 ± 2.0	50.6 ± 2.5	51.0 ± 3.5
		512	50.1%	49.8 ± 1.6	49.3 ± 2.5	51.0 ± 5.3	50.8 ± 5.7	51.6 ± 6.2
	10	128	50.1%	50.2 ± 1.2	49.6 ± 2.5	50.9 ± 2.9	50.7 ± 3.1	51.4 ± 4.9
		512	50.1%	49.9 ± 2.0	49.3 ± 2.9	50.7 ± 5.0	50.5 ± 5.2	50.4 ± 4.7
LSTM	5	32	50.3%	50.2 ± 1.1	49.9 ± 2.0	50.6 ± 2.9	50.5 ± 3.0	51.2 ± 4.6
		128	50.4%	49.5 ± 1.5	49.2 ± 2.2	49.8 ± 2.7	49.5 ± 2.8	50.4 ± 4.0
	10	32	49.3%	49.9 ± 1.1	49.5 ± 1.8	50.2 ± 2.6	49.9 ± 2.6	50.0 ± 3.8
		128	50.1%	49.9 ± 0.8	49.6 ± 1.6	50.1 ± 1.6	50.1 ± 1.7	49.8 ± 3.4
AEnc	5	32	50.5%	50.1 ± 0.7	49.7 ± 1.7	51.2 ± 2.6	51.0 ± 2.7	52.0 ± 4.2
		128	50.3%	50.1 ± 1.0	49.8 ± 1.9	50.5 ± 2.5	50.5 ± 2.5	50.8 ± 4.2
	10	32	49.8%	50.1 ± 1.1	50.0 ± 1.8	50.8 ± 2.3	50.7 ± 2.5	50.5 ± 3.5
		128	49.8%	49.8 ± 0.7	49.5 ± 1.4	50.1 ± 1.6	50.2 ± 1.7	50.2 ± 2.3
Avgs. by trading protocol				50.0 ± 1.2	49.6 ± 2.1	50.6 ± 3.0	50.4 ± 3.2	50.8 ± 4.2

Table 6. Detailed results an all five stocks for selected configurations; in bold the final capitals resulted greater than 55% with standard deviation smaller than 5%.

Model type	Index	Model accuracy	Final capital in thous by trading algorithm, initial = 50				
			SD	SDD	DB	DBD	BB
FFN (128)	CSI300	46.6%	50.2 ± 0.3	50.4 ± 0.6	51.7 ± 1.0	51.9 ± 1.2	**55.1 ± 3.6**
	HangSeng	50.1%	49.6 ± 1.7	47.3 ± 4.7	49.7 ± 2.0	48.8 ± 3.4	49.3 ± 3.3
	Nifty50	50.5%	50.1 ± 0.2	50.0 ± 0.5	50.3 ± 0.5	50.2 ± 0.5	50.8 ± 1.3
	Nikkei225	51.6%	51.4 ± 1.8	52.0 ± 2.5	52.7 ± 3.0	52.9 ± 3.2	53.0 ± 3.6
	S& P500	47.3%	49.9 ± 0.1	49.9 ± 0.2	50.3 ± 1.1	50.2 ± 1.1	48.2 ± 2.3
LSTM (32)	CSI300	46.7%	50.2 ± 0.4	50.7 ± 0.6	49.9 ± 4.0	50.6 ± 3.8	54.1 ± 5.8
	HangSeng	53.9%	49.9 ± 1.9	48.4 ± 3.4	49.8 ± 2.2	49.3 ± 2.5	49.6 ± 3.5
	Nifty50	53.4%	50.5 ± 0.6	49.9 ± 1.0	52.2 ± 3.0	51.7 ± 2.7	53.3 ± 3.6
	Nikkei225	50.9%	51.3 ± 1.7	51.5 ± 2.5	53.1 ± 3.3	53.4 ± 3.8	**54.7 ± 4.3**
	S& P500	46.6%	49.8 ± 0.1	49.7 ± 0.3	48.9 ± 0.9	48.8 ± 0.8	46.2 ± 3.3
AEnc (32)	CSI300	47.2%	50.5 ± 0.3	51.2 ± 0.5	53.2 ± 3.0	53.8 ± 2.8	**58.4 ± 4.6**
	HangSeng	51.1%	49.7 ± 1.0	48.3 ± 2.3	50.8 ± 2.7	49.4 ± 2.4	49.5 ± 2.1
	Nifty50	50.6%	50.3 ± 0.2	49.8 ± 1.0	50.1 ± 1.4	50.0 ± 1.6	51.5 ± 1.2
	Nikkei225	49.6%	50.2 ± 1.6	49.4 ± 2.9	52.7 ± 4.1	52.4 ± 4.2	52.3 ± 4.7
	S& P500	53.7%	50.0 ± 0.0	50.0 ± 0.0	50.0 ± 0.0	50.0 ± 0.0	50.1 ± 0.2
Avgs. by trading protocol			50.1 ± 1.0	49.8 ± 2.0	50.9 ± 2.5	50.7 ± 2.7	51.4 ± 4.1

and the smallest considered base size for each network (128 for FFN and 32 for LSTM and autoencoders). We can see that some stocks are generally slightly profitable than others, such as the capital values for CSI300 and Nikkei225 in bold in the last column corresponding to the BB trading protocol.

Finally, using the same data and periods, we evaluated the prediction capability of the composite autoencoder in the univariate case of the DJIA time series restricted to only the open prices. The experiment consisted in predicting the next day open price of DJIA using only its open prices of preceding 10 days. The prediction accuracy of the open price movements in the test set period was on average 64.49% over 5 runs. The profit was 63.7 ± 3.2, corresponding to 27.4% of annual gain, which is close to the maximum gain achieved by the experiments that consider only OHLC variables.

5   Conclusion

In this work, we have analyzed the predictive capabilities on the stock market of different neural network architectures, including those that in several research areas have shown superior capabilities of detecting long term dependencies in sequences of data, such as in speech recognition and in text understanding. The aim of the paper was to move away from the latest complex trends, in terms of stock market prediction based on the use of non-structured data, such as tweets, financial news, etc., in order to focus more simply on the stock time series.

From this viewpoint the work follows the philosophy of the ARIMA approach proposed in 1970 by Box and Jenkins, but experimenting advanced approaches. We tested both traditional architectures such as feed forward and LSTM networks and also a more sophisticated autoencoder-based architecture, all with the goal of daily predicting the variation of a number of target stocks. Each trained model has been evaluated through experiments, based on alternative trading protocols performing different actions in response to the network output predictions.

Our tests show that, just using historical OHLC data for a given stock, we are able to obtain predictive models which, paired to an aggressive buy/sell strategy, are able to generate a yearly profit ranging from 8.6% to 36%. This is obtained even with relatively simple and well-established neural network models, which can be trained in few minutes or hours on ordinary PCs.

The work can be extended to a scenario of parallel trading with multiple stocks at the same time, also investigating and exploiting possible correlations among different market indexes. Another possible improvement is to predict the best trading action according to forecasts of stock price movements referred to two or more days in the future. This strategy may lead to the identification of additional recurring patterns able to exploit the time lag in the reactivity of stock market as supposed in [36].

References

1. Abe, M., Nakayama, H.: Deep learning for forecasting stock returns in the cross-section. arXiv preprint arXiv:1801.01777 (2018)
2. Adebiyi, A.A., Adewumi, A.O., Ayo, C.K.: Comparison of ARIMA and artificial neural networks models for stock price prediction. J. Appl. Math. **2014**, 614342:1–614342:7 (2014). https://doi.org/10.1155/2014/614342
3. Akita, R., Yoshihara, A., Matsubara, T., Uehara, K.: Deep learning for stock prediction using numerical and textual information. In: 2016 IEEE/ACIS 15th International Conference on Computer and Information Science (ICIS), pp. 1–6. IEEE (2016)
4. Atsalakis, G.S., Valavanis, K.P.: Forecasting stock market short-term trends using a neuro-fuzzy based methodology. Expert Syst. Appl. **36**(7), 10696–10707 (2009)
5. Atsalakis, G.S., Valavanis, K.P.: Surveying stock market forecasting techniques-Part II: soft computing methods. Expert Syst. Appl. **36**(3), 5932–5941 (2009)
6. Bao, W., Yue, J., Rao, Y.: A deep learning framework for financial time series using stacked autoencoders and long-short term memory. PLOS One **12**(7), 1–24 (2017). https://doi.org/10.1371/journal.pone.0180944
7. Bollen, J., Mao, H., Zeng, X.: Twitter mood predicts the stock market. J. Comput. Sci. **2**(1), 1–8 (2011)
8. Box, G.E.P., Jenkins, G.: Time Series Analysis, Forecasting and Control. Holden-Day, Inc., San Francisco (1970)
9. Cao, Q., Leggio, K.B., Schniederjans, M.J.: A comparison between Fama and French's model and artificial neural networks in predicting the Chinese stock market. Comput. Oper. Res. **32**(10), 2499–2512 (2005)
10. Cerroni, W., Moro, G., Pasolini, R., Ramilli, M.: Decentralized detection of network attacks through P2P data clustering of SNMP data. Comput. Secur. **52**, 1–16 (2015). https://doi.org/10.1016/j.cose.2015.03.006
11. Cerroni, W., Moro, G., Pirini, T., Ramilli, M.: Peer-to-peer data mining classifiers for decentralized detection of network attacks. In: Wang, H., Zhang, R. (eds.) Proceedings of the 24th Australasian Database Conference, ADC 2013. CRPIT, vol. 137, pp. 101–108. Australian Computer Society, Inc., Darlinghurst (2013). http://crpit.com/abstracts/CRPITV137Cerroni.html
12. Cho, K., et al.: Learning phrase representations using RNN encoder-decoder for statistical machine translation. In: Proceedings of the 2014 Conference on Empirical Methods in Natural Language Processing (EMNLP), October 2014, pp. 1724–1734. Association for Computational Linguistics, Doha (2014). http://www.aclweb.org/anthology/D14-1179
13. Ding, X., Zhang, Y., Liu, T., Duan, J.: Deep learning for event-driven stock prediction. In: IJCAI, pp. 2327–2333 (2015)
14. Domeniconi, G., Masseroli, M., Moro, G., Pinoli, P.: Discovering new gene functionalities from random perturbations of known gene ontological annotations. In: KDIR 2014 - Proceedings of the International Conference on Knowledge Discovery and Information Retrieval, Rome, Italy, 21–24 October 2014, pp. 107–116. SciTePress (2014). https://doi.org/10.5220/0005087801070116
15. Domeniconi, G., Masseroli, M., Moro, G., Pinoli, P.: Cross-organism learning method to discover new gene functionalities. Comput. Methods Programs Biomed. **126**, 20–34 (2016). https://doi.org/10.1016/j.cmpb.2015.12.002

16. Domeniconi, G., Moro, G., Pagliarani, A., Pasolini, R.: Cross-domain sentiment classification via polarity-driven state transitions in a Markov model. In: Fred, A., Dietz, J.L.G., Aveiro, D., Liu, K., Filipe, J. (eds.) IC3K 2015. CCIS, vol. 631, pp. 118–138. Springer, Cham (2016). https://doi.org/10.1007/978-3-319-52758-1_8

17. Domeniconi, G., Moro, G., Pagliarani, A., Pasolini, R.: Markov chain based method for in-domain and cross-domain sentiment classification. In: KDIR 2015 - Proceedings of the International Conference on Knowledge Discovery and Information Retrieval, part of the 7th International Joint Conference on Knowledge Discovery, Knowledge Engineering and Knowledge Management (IC3K 2015), Lisbon, Portugal, vol. 1, pp. 127–137. SciTePress (2015). https://doi.org/10.5220/0005636001270137

18. Domeniconi, G., Moro, G., Pagliarani, A., Pasolini, R.: Learning to predict the stock market Dow Jones index detecting and mining relevant tweets. In: Proceedings of the 9th International Joint Conference on Knowledge Discovery, Knowledge Engineering and Knowledge Management, Funchal, Madeira, Portugal, 1–3 November 2017, vol. 1, pp. 165–172. SciTePress (2017). https://doi.org/10.5220/0006488201650172

19. Domeniconi, G., Moro, G., Pagliarani, A., Pasolini, R.: On deep learning in cross-domain sentiment classification. In: Proceedings of the 9th International Joint Conference on Knowledge Discovery, Knowledge Engineering and Knowledge Management, Funchal, Madeira, Portugal, 1–3 November 2017, vol. 1, pp. 50–60 (2017). https://doi.org/10.5220/0006488100500060

20. Domeniconi, G., Moro, G., Pasolini, R., Sartori, C.: Cross-domain text classification through iterative refining of target categories representations. In: KDIR 2014 - Proceedings of the International Conference on Knowledge Discovery and Information Retrieval, Rome, Italy, 21–24 October 2014, pp. 31–42. SciTePress (2014). https://doi.org/10.5220/0005069400310042

21. Domeniconi, G., Moro, G., Pasolini, R., Sartori, C.: Iterative refining of category profiles for nearest centroid cross-domain text classification. In: Fred, A., Dietz, J.L.G., Aveiro, D., Liu, K., Filipe, J. (eds.) IC3K 2014. CCIS, vol. 553, pp. 50–67. Springer, Cham (2015). https://doi.org/10.1007/978-3-319-25840-9_4

22. Domeniconi, G., Moro, G., Pasolini, R., Sartori, C.: A comparison of term weighting schemes for text classification and sentiment analysis with a supervised variant of tf.idf. In: Helfert, M., Hölzinger, A., Belo, O., Francalanci, C. (eds.) DATA 2015. CCIS, vol. 584, pp. 39–58. Springer, Cham (2016). https://doi.org/10.1007/978-3-319-30162-4_4

23. Esteva, A., et al.: Dermatologist-level classification of skin cancer with deep neural networks. Nature **542**(7639), 115–118 (2017). https://doi.org/10.1038/nature21056

24. Fabbri, M., Moro, G.: Dow Jones trading with deep learning: the unreasonable effectiveness of recurrent neural networks. In: Proceedings of the 7th International Conference on Data Science, Technology and Applications - Volume 1: DATA, pp. 142–153. INSTICC, SciTePress (2018). https://doi.org/10.5220/0006922101420153

25. Fama, E.F.: Efficient capital markets: a review of theory and empirical work. J. Financ. **25**(2), 383–417 (1970)

26. Fischer, T., Krauss, C.: Deep learning with long short-term memory networks for financial market predictions. Eur. J. Oper. Res. **270**, 654–669 (2017)

27. Fisher, I.E., Garnsey, M.R., Hughes, M.E.: Natural language processing in accounting, auditing and finance: a synthesis of the literature with a roadmap for future research. Intell. Syst. Account. Financ. Manag. **23**(3), 157–214 (2016)

28. Gers, F.A., Schmidhuber, J., Cummins, F.: Learning to forget: continual prediction with LSTM. Neural Comput. **12**, 2451–2471 (1999)
29. Gidofalvi, G., Elkan, C.: Using news articles to predict stock price movements. Department of Computer Science and Engineering, University of California, San Diego (2001)
30. Graves, A.: Generating sequences with recurrent neural networks. arXiv preprint arXiv:1308.0850 (2013)
31. Hochreiter, S., Schmidhuber, J.: Long short-term memory. Neural Comput. **9**(8), 1735–1780 (1997)
32. Khashei, M., Bijari, M., Ardali, G.A.R.: Improvement of auto-regressive integrated moving average models using fuzzy logic and artificial neural networks (ANNs). Neurocomputing **72**(4–6), 956–967 (2009)
33. Khashei, M., Bijari, M., Ardali, G.A.R.: Hybridization of autoregressive integrated moving average (ARIMA) with probabilistic neural networks (PNNs). Comput. Ind. Eng. **63**(1), 37–45 (2012)
34. Kingma, D.P., Ba, J.: Adam: a method for stochastic optimization. CoRR abs/1412.6980 (2014). http://arxiv.org/abs/1412.6980
35. Krizhevsky, A., Sutskever, I., Hinton, G.E.: ImageNet classification with deep convolutional neural networks. In: Proceedings of the 25th International Conference on Neural Information Processing Systems, NIPS 2012, vol. 1, pp. 1097–1105. Curran Associates Inc., USA (2012). http://dl.acm.org/citation.cfm?id=2999134.2999257
36. LeBaron, B., Arthur, W.B., Palmer, R.: Time series properties of an artificial stock market. J. Econ. Dyn. Control. **23**(9–10), 1487–1516 (1999)
37. Lee, C.M., Ko, C.N.: Short-term load forecasting using lifting scheme and ARIMA models. Expert Syst. Appl. **38**(5), 5902–5911 (2011)
38. di Lena, P., Domeniconi, G., Margara, L., Moro, G.: GOTA: GO term annotation of biomedical literature. BMC Bioinform. **16**, 346:1–346:13 (2015). https://doi.org/10.1186/s12859-015-0777-8
39. Lin, M.C., Lee, A.J., Kao, R.T., Chen, K.T.: Stock price movement prediction using representative prototypes of financial reports. ACM Trans. Manag. Inf. Syst. (TMIS) **2**(3), 19 (2011)
40. Lo, A.W., MacKinlay, A.C.: Stock market prices do not follow random walks: evidence from a simple specification test. Rev. Financ. Stud. **1**(1), 41–66 (1988). https://doi.org/10.1093/rfs/1.1.41
41. Lo, A.W., Repin, D.V.: The psychophysiology of real-time financial risk processing. J. Cogn. Neurosci. **14**(3), 323–339 (2002)
42. Lodi, S., Monti, G., Moro, G., Sartori, C.: Peer-to-peer data clustering in self-organizing sensor networks. In: Intelligent Techniques for Warehousing and Mining Sensor Network Data, December 2009, pp. 179–211. IGI Global, Information Science Reference, Hershey (2009). http://www.igi-global.com/chapter/peer-peer-data-clustering-self/39546
43. Malkiel, B.G.: The efficient market hypothesis and its critics. J. Econ. Perspect. **17**(1), 59–82 (2003)
44. Malkiel, B.G.: A Random Walk Down Wall Street. Norton, New York (1973)
45. Merh, N., Saxena, V.P., Pardasani, K.R.: A comparison between hybrid approaches of ann and arima for indian stock trend forecasting. Bus. Intell. J. **3**(2), 23–43 (2010)
46. Mitra, S.K.: Optimal combination of trading rules using neural networks. Int. Bus. Res. **2**(1), 86 (2009)

47. Monti, G., Moro, G.: Self-organization and local learning methods for improving the applicability and efficiency of data-centric sensor networks. In: Bartolini, N., Nikoletseas, S., Sinha, P., Cardellini, V., Mahanti, A. (eds.) QShine 2009. LNIC-SSITE, vol. 22, pp. 627–643. Springer, Heidelberg (2009). https://doi.org/10.1007/978-3-642-10625-5_40

48. Moro, G., Pagliarani, A., Pasolini, R., Sartori, C.: Cross-domain & in-domain sentiment analysis with memory-based deep neural networks. In: Proceedings of the 10th International Joint Conference on Knowledge Discovery, Knowledge Engineering and Knowledge Management - Volume 1: KDIR, pp. 127–138. INSTICC, SciTePress (2018). https://doi.org/10.5220/0007239101270138

49. Moro, G., Pasolini, R., Domeniconi, G., Pagliarani, A., Roli, A.: Prediction and trading of Dow Jones from Twitter: a boosting text mining method with relevant tweets identification. In: Fred, A., Aveiro, D., Dietz, J.L.G., Liu, K., Bernardino, J., Salgado, A., Filipe, J. (eds.) IC3K 2017. CCIS, vol. 976, pp. 26–42. Springer, Cham (2019). https://doi.org/10.1007/978-3-030-15640-4_2

50. Mostafa, M.M.: Forecasting stock exchange movements using neural networks: empirical evidence from Kuwait. Expert Syst. Appl. **37**(9), 6302–6309 (2010)

51. Olson, D., Mossman, C.: Neural network forecasts of Canadian stock returns using accounting ratios. Int. J. Forecast. **19**(3), 453–465 (2003)

52. Pagliarani, A., Moro, G., Pasolini, R., Domeniconi, G.: Transfer learning in sentiment classification with deep neural networks. In: Fred, A., Aveiro, D., Dietz, J.L.G., Liu, K., Bernardino, J., Salgado, A., Filipe, J. (eds.) IC3K 2017. CCIS, vol. 976, pp. 3–25. Springer, Cham (2019). https://doi.org/10.1007/978-3-030-15640-4_1

53. Schumaker, R., Chen, H.: Textual analysis of stock market prediction using financial news articles. In: AMCIS 2006 Proceedings, p. 185 (2006)

54. Schumaker, R.P., Chen, H.: Textual analysis of stock market prediction using breaking financial news: the AZFin text system. ACM Trans. Inf. Syst. (TOIS) **27**(2), 12 (2009)

55. Soni, S.: Applications of ANNs in stock market prediction: a survey. Int. J. Comput. Sci. Eng. Technol. **2**(3), 71–83 (2011)

56. Srivastava, N., Mansimov, E., Salakhutdinov, R.: Unsupervised learning of video representations using LSTMs. CoRR abs/1502.04681 (2015). http://arxiv.org/abs/1502.04681

57. Sterba, J., Hilovska, K.: The implementation of hybrid ARIMA neural network prediction model for aggregate water consumption prediction. Aplimat J. Appl. Math. **3**(3), 123–131 (2010)

58. Tetlock, P.C.: Giving content to investor sentiment: the role of media in the stock market. J. Financ. **62**(3), 1139–1168 (2007)

59. Wang, W., Li, Y., Huang, Y., Liu, H., Zhang, T.: A method for identifying the mood states of social network users based on cyber psychometrics. Future Internet **9**(2), 22 (2017)

60. Wikipedia Contributors: Financial statement – Wikipedia, the free encyclopedia (2018). https://en.wikipedia.org/w/index.php?title=Financial_statement&oldid=831492885. Accessed 21 Feb 2019

Author Index

Printed in the United States
By Bookmasters